D0464855

ALSO BY GREGORY CHAITIN

Algorithmic Information Theory

Conversations with a Mathematician

Exploring Randomness

From Philosophy to Program Size

Information, Randomness, and Incompleteness

Information-Theoretic Incompleteness

The Limits of Mathematics

The Unknowable

META MATH!

META MATH!

THE QUEST FOR OMEGA

Gregory Chaitin

A Peter N. Névraumont Book

PANTHEON BOOKS, NEW YORK

 Published in the United States by Pantheon Books, a division of Random House, Inc., New York, and simultaneously in Canada by Random House of Canada Limited, Toronto.

Pantheon Books and colophon are registered trademarks of Random House, Inc.

Portions of this book were originally published in *American Scientist* and by Akademie Verlag, Berlin.

Owing to limitations of space, all acknowledgments for permission to reprint previously published material may be found preceding the author's biography.

Library of Congress Cataloging-in-Publication Data

Chaitin, Gregory J.
Meta math! : the quest for omega / Gregory Chaitin.
p. cm.
Includes bibliographical references and index.
ISBN 0-375-42313-3
1. Machine theory. 2. Computational complexity. 3. Stochastic processes.
I. Title.

QA276.C435 2005 510—dc22 2005043044

www.pantheonbooks.com

Printed in the United States of America
First Edition
2 4 6 8 9 7 5 3 1

WILLIAM BLAKE:
The Ancient of Days, 1794.

À la femme de mes rêves, the eternal muse,
and the search for inexpressible beauty . . .

CONTENTS

PREFACE

Science is an open road: each question you answer raises **ten** new questions, and much more difficult ones! Yes, I'll tell you some things I discovered, but the journey is endless, and mostly I'll share with you, the reader, my doubts and preoccupations and what I think are promising and challenging new things to think about.

It would be easy to spend many lifetimes working on any of a number of the questions that I'll discuss. That is how the good questions are. You can't answer them in five minutes, and it would be no good if you could.

Science is an adventure. I don't believe in spending years studying the work of others, years learning a complicated field before I can contribute a tiny little bit. I prefer to stride off in totally new directions, where imagination is, at least initially, much more important than technique, because the techniques have yet to be developed. It takes all kinds of people to advance knowledge, the pioneers, and those who come afterwards and patiently work a farm. This book is for pioneers!

Most books emphasize what the author knows. I'll try to emphasize what I would like to know, what I'm hoping someone will discover, and just how much there is that is fundamental and that we **don't know!**

And yes, I'm a mathematician, but I'm really interested in everything: what is life, what's intelligence, what is consciousness, does the universe contain randomness, are space and time continuous or discrete. To me math is just the fundamental tool of philosophy, it's a way to work out ideas, to flesh them out, to build models, **to understand**! As Leibniz said, without math you cannot really understand philosophy, without philosophy you cannot really understand mathematics, and with neither of them, you can't really understand a thing! Or at least that's my credo, that's how I operate.

On the other hand, as someone said a long time ago, "a mathematician who is not something of a poet will never be a good mathematician." And "there is no permanent place in the world for ugly mathematics" (G. H. Hardy). To survive, mathematical ideas must be beautiful, they must be seductive, and they must be illuminating, they must help us to understand, they must inspire us. So I hope this little book will also convey something of this more personal aspect of mathematical creation, of mathematics as a way of celebrating the universe, as a kind of love-making! I want you to fall in love with mathematical ideas, to begin to feel seduced by them, to see how easy it is to be entranced and to want to spend years in their company, years working on mathematical projects.

And it is a mistake to think that a mathematical idea can survive merely because it is **useful,** because it has practical applications. On the contrary, what is useful varies as a function of time, while "a thing of beauty is a joy forever" (Keats). Deep theory is what is really useful, not the ephemeral usefulness of practical applications!

Part of that beauty, an essential part, is the clarity and sharpness that the mathematical way of thinking about things promotes and achieves. Yes, there are also mystic and poetic ways of relating to the world, and to create a new math theory, or to discover new mathematics, you have to feel comfortable with vague, unformed, embryonic ideas, even as you try to sharpen them. But one of the things about math that seduced me as a child was the black/whiteness, the clarity and sharpness of the world of mathematical ideas, that is so different from the messy (but wonderful!) world of human emotions and interpersonal complications! No wonder that scientists express their understanding in mathematical terms, when they can!

As has been often said, to understand something is to make it mathematical, and I hope that this may eventually even happen to the fields of psychology and sociology, someday. That is my bias, that the math point of view can contribute to everything, that it can help to clarify anything. Mathematics is a way of characterizing or expressing **structure.** And the universe seems to be built, at some fundamental level, out of mathematical structure. To speak metaphorically, it appears that God is a mathematician, and that the structure of the world—God's thoughts!—are mathematical, that this is the cloth out of which the world is woven, the wood out of which the world is built . . .

When I was a child the excitement of relativity theory (Einstein!) and quantum mechanics was trailing off, and the excitement of DNA and molecular biology had not begun. What was the new big thing then? The Computer! Which some people referred to then as "Giant Electronic Brains." I was fascinated by computers as a child. First of all, because they were a great toy, an infinitely malleable artistic medium of creation. I loved programming! But most of all, because the computer was (and still is!) a wonderful new **philosophical** and mathematical concept. The computer is even more revolutionary as an **idea,** than it is as a practical device that alters society—and we all know how much it has changed our lives. Why do I say this? Well, the computer changes epistemology, it changes the meaning of "to understand." To me, you understand something only if you can program it. (You, not someone else!) Otherwise you don't really understand it, you only **think** you understand it.

And, as we shall see, the computer changes the way you do mathematics, it changes the kind of mathematical models of the world that you build. In a nutshell, now God seems to be a programmer, not a mathematician! The computer has provoked a paradigm shift: it suggests a digital philosophy, it suggests a new way of looking at the world, in which everything is discrete and nothing is continuous, in which everything is digital information, 0's and 1's. So I was naturally attracted to this revolutionary new idea.

And what about the so-called real world of money and taxes and disease and death and war? What about this "best of all possible worlds, in which everything is a necessary evil"!? Well, I prefer to ignore that insignificant world and concentrate instead on the world of ideas, on the quest for understanding. Instead of looking down into the mud, how about looking up at the stars! Why don't you try it? Maybe you'll like that kind of a life too!

Anyway, you don't have to read these musings from cover to cover. Just leap right in wherever you like, and on a first reading please skip anything that seems too difficult. Maybe it'll turn out afterwards that it's actually not so difficult . . . I think that basic ideas are simple. And I'm not really interested in complicated ideas, I'm only interested in fundamental ideas. If the answer is extremely complicated, I think that probably means that we've asked the wrong question!

No man is an island, and practically every page of this book has benefited from discussions with Françoise Chaitin-Chatelin during the past decade; she has wanted me to write this book for that long. Gradiva, a Portuguese publishing house, arranged a stimulating visit for me to that country in January 2004. During that trip I gave a talk at the University of Lisbon on Chapter Five of this book.

I am also grateful to Jorge Aguirre for inviting me to present this book as a course at his summer school at the University of Río Cuarto in Córdoba, Argentina, in February 2004; the students' comments there were extremely helpful. And Cristian Calude has provided me with a delightful environment in which to finish this book at his Center for Discrete Math and Theoretical Computer Science at the University of Auckland.

Finally, I am greatly indebted to Nabil Amer and to the Physics Department at the IBM Thomas J. Watson Research Center in Yorktown Heights, New York (my home base between trips) for their support of my work.

—GREGORY CHAITIN
Auckland, New Zealand,
March 2004

META MATH!

QUOTES BY LEIBNIZ/GALILEO

Sans les mathématiques on ne pénètre point au fond de la philosophie.
Sans la philosophie on ne pénètre point au fond des mathématiques.
Sans les deux on ne pénètre au fond de rien.

Without mathematics we cannot penetrate deeply into philosophy.
Without philosophy we cannot penetrate deeply into mathematics.
Without both we cannot penetrate deeply into anything.

—LEIBNIZ

La filosofia è scritta in questo grandissimo libro
che continuamente ci sta aperto innanzi a gli occhi (io dico l'universo)
ma non si può intender se prima non s'impara a intender
la lingua e conoscere i caratteri ne' quali è scritto.
Egli è scritto in lingua matematica e
i caratteri sono triangoli, cerchi, ed altre figure geometriche
senza i quali mezi è impossibile a intenderne umanamente parola;
senza questi è un aggirarsi vanamente per un oscuro laberinto.

Philosophy is written in this very great book
which always lies open before our eyes (I mean the universe),
but one cannot understand it unless one first learns to understand
the language and recognize the characters in which it is written.
It is written in mathematical language and
the characters are triangles, circles and other geometrical figures;
without these means it is humanly impossible to understand a word of it;
without these there is only clueless scrabbling around in a dark labyrinth.

—GALILEO

FRANZ KAFKA: BEFORE THE LAW

Before the law sits a gatekeeper. To this gatekeeper comes a man from the country who asks to gain entry into the law. But the gatekeeper says that he cannot grant him entry at the moment. The man thinks about it and then asks if he will be allowed to come in later on. "It is possible," says the gatekeeper, "but not now." At the moment the gate to the law stands open, as always, and the gatekeeper walks to the side, so the man bends over in order to see through the gate into the inside. When the gatekeeper notices that, he laughs and says: "If it tempts you so much, try it in spite of my prohibition. But take note: I am powerful. And I am only the most lowly gatekeeper. But from room to room stand gatekeepers, each more powerful than the other. I can't even endure even one glimpse of the third." The man from the country has not expected such difficulties: the law should always be accessible for everyone, he thinks, but as he now looks more closely at the gatekeeper in his fur coat, at his large pointed nose and his long, thin, black Tartar's beard, he decides that it would be better to wait until he gets permission to go inside. The gatekeeper gives him a stool and allows him to sit down at the side in front of the gate. There he sits for days and years. He makes many attempts to be let in, and he wears the gatekeeper out with his requests. The gatekeeper often interrogates him briefly, questioning him about his homeland and many other things, but they are indifferent questions, the kind great men put, and at the end he always tells him once more that he cannot let him inside yet. The man, who has equipped himself with many things for his journey, spends everything, no matter how valuable, to win over the gatekeeper. The latter takes it all but, as he does so, says, "I am taking this only so that you do not think you have failed to do anything." During the many years the man observes the gatekeeper almost continuously. He forgets the other gatekeepers, and this one seems to him the only obstacle for entry into the law. He curses the unlucky circumstance, in the first years thoughtlessly and out loud, later, as he grows old, he still mumbles to himself. He becomes childish and, since in the long years studying the gatekeeper he has come to know the fleas in his fur collar, he even asks the fleas to help him persuade the gatekeeper. Finally his eyesight grows weak, and he does not know whether things are really darker around him or whether his eyes are merely deceiving him. But he recognizes now in the darkness an illumination which breaks inextinguishably out of the gateway to the law. Now he no longer has much time to live. Before his death he gathers in his head all his experiences of the entire time up into one question which he has not yet put to the gatekeeper. He waves to him, since he can no longer lift up his stiffening body. The gatekeeper has to bend way down to him, for the great difference has changed things to the disadvantage of the man. "What do you still want to know, then?" asks the gatekeeper. "You are insatiable." "Everyone strives after the law," says the man, "so how is it that in these many years no one except me has requested entry?" The gatekeeper sees that the man is already dying and, in order to reach his diminishing sense of hearing, he shouts at him, "Here no one else can gain entry, since this entrance was assigned only to you. I'm going now to close it."

[This translation is by Ian Johnston of Malaspina College, Canada. Note that in Hebrew "Law" is "Torah," which also means "Truth." Orson Welles delivers a beautiful reading of this parable at the very beginning of his film version of Kafka's *The Trial.*]

One

INTRODUCTION

In his book *Everything and More: A Compact History of Infinity,* David Foster Wallace refers to Gödel as "modern math's absolute Prince of Darkness" (p. 275) and states that because of him "pure math's been in mid-air for the last 70 years" (p. 284). In other words, according to Wallace, since Gödel published his famous paper in 1931, mathematics has been suspended hanging in mid-air without anything like a proper foundation.

It is high time these dark thoughts were permanently laid to rest. Hilbert's century-old vision of a static completely mechanical absolutely rigorous formal mathematics was a misguided attempt intended to demonstrate the absolute certainty of mathematical reasoning. It is time for us to recover from this disease!

Gödel's 1931 work on incompleteness, Turing's 1936 work on uncomputability, and my own work on the role of information, randomness and complexity have shown increasingly emphatically that the role that Hilbert envisioned for formalism in mathematics is best served by computer programming languages, which **are** in fact formalisms that can be mechanically interpreted—but they are formalisms for computing and calculating, not for reasoning, not for proving theorems, and most emphatically not for inventing new mathematical concepts nor for making new mathematical discoveries.

In my opinion, the view that math provides absolute certainty and is static and perfect while physics is tentative and constantly evolving is a false dichotomy. Math is actually not that different from physics. Both are attempts of the human mind to organize, to make sense of, human experience; in the case of physics, experience in the laboratory, in the physical world; and in the case of math, experience in the computer, in the mental mindscape of pure mathematics.

And mathematics is far from static and perfect; it is constantly evolving, constantly changing, constantly morphing itself into new forms. New concepts are constantly transforming math and creating new fields, new viewpoints, new emphasis, and new questions to answer. And mathematicians do in fact utilize unproved new principles suggested by computational experience, just as a physicist would.

And in discovering and creating new mathematics, mathematicians do base themselves on intuition and inspiration, on unconscious motivations and impulses, and on their aesthetic sense, just like any creative artist would. And mathematicians do not lead logical mechanical "rational" lives. Like any creative artist, they are passionate emotional people who deeply care about their art, they are unconventional eccentrics motivated by mysterious forces, not by money nor by a concern for the "practical applications" of their work.

I know, because I'm one of these crazy people myself! I've been obsessed by these questions for my whole life, starting at an early age. And I'll give you an insider's view of all of this, a firsthand report from the front, where there is still a lot of fighting, a lot of pushing and shoving, between different viewpoints. In fact basic questions like this are never settled, never definitively put aside, they have a way of resurfacing, of popping up again in transformed form, every few generations . . .

So that's what this book is about: It's about reasoning questioning itself, and its limits and the role of creativity and intuition, and the sources of new ideas and of new knowledge. That's a big subject, and I only understand a little bit of it, the areas that I've worked in or experienced myself. Some of this **nobody** understands very well, it's a task for the future. How about **you**?! Maybe you can do some work in this area. Maybe you can push the darkness back a millimeter or two! Maybe you can come up with an important new idea, maybe you can imagine a new kind of question to ask, maybe you can transform the landscape by seeing it from a different point of view! That's all it takes, just one little new idea, and lots and lots of hard work to develop it and to convince other people! Maybe you can put a scratch on the rock of eternity!

Remember that math is a free creation of the human mind, and as Cantor—the inventor of the modern theory of infinity described by Wallace—said, the essence of math resides in its freedom, in the freedom to create. But history judges these creations by their enduring beauty and by the extent to which they illuminate other mathematical ideas or the physical universe, in a word, by their "fertility." Just as the beauty of a woman's breasts or the delicious curve of her hips is actually concerned with childbearing, and isn't merely for the delight of painters and photographers, so a math idea's beauty also has something to do with its "fertility," with the extent to which it enlightens us, illuminates us, and inspires us with other ideas and suggests unsuspected connections and new viewpoints.

Anyone can define a new mathematical concept—many mathematical papers do—but only the beautiful and the fertile ones survive. It's sort of similar to what Darwin termed "sexual selection," which is the way animals (including us) choose their mates for their beauty. This is a part of Darwin's original theory of evolution that nowadays one usually doesn't hear much about, but in my opinion it is much to be preferred to the "survival of the fittest" and the "nature red in tooth and claw" views of biological evolution. As an example of this unfortunate neglect, in a beautiful edition of Darwin's *The Descent of Man* that I happen to possess, several chapters on sexual selection have been completely eliminated!

So this gives some idea of the themes that I'll be exploring with you. Now let me outline the contents of the book.

OVERVIEW OF THE BOOK

Here's our road to Ω:

- In the second chapter I'll tell you how the idea of the computer entered mathematics and quickly established its usefulness.

- In the third chapter, we'll add the idea of algorithmic information, of measuring the size of computer programs.
- The intermezzo briefly discusses physical arguments against infinite-precision real numbers.
- The fifth chapter analyzes such numbers from a mathematical point of view.
- Finally the sixth chapter presents my information-based analysis of what mathematical reasoning can or cannot achieve. Here's Ω in all its glory.
- A brief concluding chapter discusses creativity . . .
- And there's a short list of suggested books, plays, and even musicals!

Now let's get to work . . .

Two

THREE STRANGE LOVES: PRIMES/GÖDEL/LISP

I am a mathematician, and this is a book about mathematics. So I'd like to start by sharing with you my vision of mathematics: why is it beautiful, how it advances, what fascinates me about it. And in order to do this I will give some case histories. I think that it is useless to make general remarks about mathematics without exhibiting some specific examples. So you and I are going to do some real mathematics together: important mathematics, significant mathematics. I will try to make it as easy as possible, but I'm going to show you the real thing! If you can't understand something, my advice is to just skim it to get the flavor of what is going on. Then, if you are interested, come back later and try to work your way through it slowly with a pencil and paper, looking at examples and special cases. Or you can just skip all the math and read the general observations about the nature of mathematics and the mathematical enterprise that my examples are intended to illustrate!

In particular, this chapter will try to make the case the computer isn't just a billion (or is it trillion?) dollar industry, it is also—which is more important to me—an extremely significant and fundamental new concept that changes the way you think about mathematical problems. Of course, you can use a computer to check examples or to do massive calculations, but I'm not talking about that. I'm interested in the computer as a new **idea,** a new and fundamental philosophical concept that changes mathematics, that solves old problems better and suggests new problems, that changes our way of thinking and helps us to understand things better, that gives us radically new insights . . .

And I should say right away that I completely disagree with those who say that the field of mathematics embodies static eternal perfection, and that mathematical ideas are inhuman and unchanging. On the contrary, these case studies, these intellectual histories, illustrate the fact that mathematics is constantly evolving and changing, and that our perspective, even on basic and deep mathematical questions, often shifts in amazing and unexpected fashion.[1] All it takes is a new idea! You just have to be inspired, and then work like mad to develop your new viewpoint. People will fight you at first, but if you are right, then everyone will eventually say that it was **obviously** a better way to think about the problem, and that you contributed little or nothing! In a way, that is the greatest compliment. And that's exactly what happened to Galileo: he is a good example of this phenomenon in the history of ideas. The paradigm shift that he fought so hard to achieve is now taken so absolutely, totally and thoroughly for granted, that we can no longer even understand how much he actually contributed! We can no longer conceive of any other way of thinking about the problem.

And although mathematical ideas and thought are constantly evolving, you will also see that the most basic fundamental problems never go away. Many of these problems go back to the ancient Greeks, and maybe even to ancient Sumer, although we may never know for sure. The fundamental philosophical questions like the continuous versus the discrete or the limits of knowledge are **never** definitively solved. Each generation formulates its own answer, strong personalities briefly impose their views, but the feeling of satisfaction is always temporary, and then the process continues, it continues forever. Because in order to be able to fool yourself into thinking that you have solved a really fundamental problem, you have to shut your eyes and focus on only one tiny little aspect of the problem. Okay, for a while you can do that, you can and should make progress that way. But after the brief elation of "victory," you, or other people who come after you, begin to realize that the problem that you solved was only a toy version of the real problem, one that leaves out significant aspects of the problem, aspects of the problem that in fact you **had** to

[1]See also the history of proofs that there are transcendental numbers in Chapter Five.

ignore in order to be able to get anywhere. And those forgotten aspects of the problem never go away entirely: Instead they just wait patiently outside your cozy little mental construct, biding their time, knowing that at some point someone else is going to have to take them seriously, even if it takes hundreds of years for that to happen!

And finally, let me also say that I think that the history of ideas is the best way to learn mathematics. I always hated textbooks. I always hated books full of formulas, dry books with no colorful opinions, with no personality! The books I loved were books where the author's personality shows through, books with lots of words, explanations and ideas, not just formulas and equations! I still think that the best way to learn a new idea is to see its history, to see why someone was forced to go through the painful and wonderful process of giving birth to a new idea! To the person who discovered it, a new idea seems inevitable, unavoidable. The first paper may be clumsy, the first proof may not be polished, but that is raw creation for you, just as messy as making love, just as messy as giving birth! But you **will** be able to see where the new idea comes from. If a proof is "elegant," if it's the result of two hundred years of finicky polishing, it will be as inscrutable as a direct divine revelation, and it's impossible to guess how anyone could have discovered or invented it. It will give you no insight, no, probably none at all.

Enough talk! Let's begin! I'll have much more to say after we see a few examples.

AN EXAMPLE OF THE BEAUTY OF MATHEMATICS: THE STUDY OF THE PRIME NUMBERS

The primes

$$2, 3, 5, 7, 11, 13, 17, 19, 23, 29, 31, 37 \ldots$$

are the unsigned whole numbers with no exact divisors except themselves and 1. It is usually better not to consider 1 a prime, for technical reasons (so that you get unique factorization into primes, see below). So 2 is the only even prime and

$$9 = 3 \times 3,\ \ 15 = 3 \times 5,\ \ 21 = 3 \times 7,\ \ 25 = 5 \times 5,$$

$$27 = 3 \times 9,\ \ 33 = 3 \times 11,\ \ 35 = 5 \times 7 \ldots$$

are **not** primes. If you keep factorizing a number, eventually you have to reach primes, which can't be broken down any more. For example:

$$100 = 10 \times 10 = (2 \times 5) \times (2 \times 5) = 2^2 \times 5^2.$$

Alternatively,

$$100 = 4 \times 25 = (2 \times 2) \times (5 \times 5) = 2^2 \times 5^2.$$

Note that the final results are the same. That this is always the case was already demonstrated by Euclid two thousand years ago, but amazingly enough a much simpler proof was discovered recently (in the last century). To make things easier in this chapter, though, I'm not going to use the fact that this is always the case, that factorization into primes is unique. So we can go ahead without delay; we don't need to prove that.

The ancient Greeks came up with all these ideas two millennia ago, and they have fascinated mathematicians ever since. What is fascinating is that as simple as the whole numbers and the primes are, it is easy to state clear, straightforward questions about them that **nobody** knows how to answer, not even in two thousand years, not even the best mathematicians on earth!

Now I'd like to mention two ideas that we'll discuss a lot later: **irreducibility** and **randomness.**

Primes are *irreducible numbers,* irreducible via multiplication, via factoring . . .

And what's mysterious about the primes is that they seem to be scattered about in a haphazard manner. In other words, the primes exhibit some kind of *randomness,* since the local details of the distribution of primes shows no apparent order, even though we can calculate them one by one.

By the way, Chapters 3 and 4 of Stephen Wolfram's *A New Kind of Science* give many other examples of simple rules that yield extremely complicated behavior. But the primes were the first example of this phenomenon that people noticed.

For example, it is easy to find an arbitrarily long gap in the primes. To find $N - 1$ non-prime numbers in a row—these are called composite numbers—just multiply all the whole numbers from 1 to N together (that's called $N!$, N factorial), and add in turn 2, 3, 4 up through N to this product $N!$:

$$N! + 2, \quad N! + 3, \quad \ldots \quad N! + (N - 1), \quad N! + N.$$

None of these numbers are prime. In fact, the first one is divisible by 2, the second one is divisible by 3, and the last one is divisible by N. But you don't really need to go that far out.

In fact the size of the gaps in the primes seem to jump around in a fairly haphazard manner too. For example, it looks like there are infinitely many twin primes, consecutive odd primes separated by just one even number. The computational evidence is very persuasive. But no one has managed to prove this yet!

And it is easy to show that there are infinitely many primes—we'll do this three different ways below—but in whatever direction you go you quickly get to results which are conjectured, but which no one knows how to prove. So the frontiers of knowledge are nearby, in fact, extremely close.

For example, consider "perfect" numbers like 6, which is equal to the sum of all its proper divisors—divisors less than the number itself—since $6 = 3 + 2 + 1$. Nobody knows if there are infinitely many perfect numbers. It is known that each Mersenne prime, each prime of the form $2^n - 1$, gives you a perfect number, namely $2^{n-1} \times (2^n - 1)$, and that there are many Mersenne primes and that every even perfect number is generated from a Mersenne prime in this way. But no one knows if there are infinitely many Mersenne primes. And nobody knows if there are **any** odd perfect numbers. Nobody has ever seen one, and they would have to be very large, but no one is sure what's going on here . . .[2]

So right away you get to the frontier, to questions that nobody knows how to answer. Nevertheless many young math students, children and teenagers who have been bitten by the math bug, work away on these problems, hoping that they may succeed where everyone else has failed. Which indeed they may! Or at least they may find something interesting along the way, even if they don't get all the way to their goal. Sometimes a fresh look is better, sometimes it's better not to know what everyone else has done, especially if they were all going in the wrong direction anyway!

For example, there is the recent discovery of a fast algorithm to check if a number is prime by two students and their professor in India, a simple algorithm that had been missed by all the experts. My friend Professor Jacob Schwartz at the Courant Institute of New York University even had the idea of including a few famous unsolved math problems on his final exams, in the hope that a brilliant student who was not aware that they were famous problems might in fact manage to solve one of them!

Yes, miracles can happen. But they don't happen that often. And the question of whether we are burdened by our current knowledge is in fact a serious one.

Are the primes the right concept? How about perfect numbers? A concept is only as good as the theorems that it leads to! Perhaps we have been following the wrong clues. How much of our current mathematics is habit, and how much is essential? For example, instead of primes, perhaps we should be concerned with the opposite, with "maximally divisible numbers"! In fact the brilliantly intuitive mathematician Ramanujan, who was self-taught, and Doug Lenat's artificial intelligence program AM (for Automated Mathematician) both came up with just such a concept. Would mathematics done on another planet by intelligent aliens be similar to or very different from ours?

---

[2]For more on this subject, see Tobias Dantzig's beautiful history of ideas, *Number, The Language of Science.*

As the great French mathematician Henri Poincaré said, "Il y a les problèmes que l'on se pose, et les problèmes qui se posent." There are questions that one asks, and questions that ask themselves! So just how inevitable are our current concepts? If evolution were rerun, would humans reappear? If the history of math were rerun, would the primes reappear? Not sure! Wolfram has examined this question in Chapter 12 of his book, and he comes up with interesting examples that suggest that our current mathematics is much more arbitrary than most people think.

In fact the aliens are right here on this very planet! Those amazing Australian animals, and the mathematicians of previous centuries, they are the aliens, and they are very different from us. Mathematical style and fashion varies substantially as a function of time, even in the comparatively recent past . . .

Anyway, here comes our first case study of the history of ideas in mathematics. Why are there infinitely many primes?

EUCLID'S PROOF THAT THERE ARE INFINITELY MANY PRIMES

We'll prove that there are infinitely many primes by assuming that there are only finitely many and deriving a contradiction. This is a common strategy in mathematical proofs, and it's called a proof by *reductio ad absurdum,* which is Latin for "reduction to an absurdity."

So let's suppose the opposite of what we wish to prove, namely that there are only finitely many primes, and in fact that K is the very last prime number. Now consider

$$1 + K! = 1 + (1 \times 2 \times 3 \times \ldots \times K).$$

This is the product of all the positive integers up to what we've assumed is the very last prime, plus one. But when this number is divided by any prime, it leaves the remainder 1! So it must itself be a prime! Contradiction! So our initial assumption that K was the largest prime has got to be false.

Here's another way to put it. Suppose that all the primes that we know are less than or equal to N. How can we show that there has to be a bigger prime? Well, consider $N! + 1$. Factorize it completely, until you get only primes. Each of these primes has to be greater than N, because no number $\leq N$ divides $N! + 1$ exactly. So the next prime greater than N has to be $\leq N! + 1$.

This masterpiece of a proof has never been equalled, even though it is 2000 years old! The balance between the ends and the means is quite remarkable. And it shows that people were just as intelligent 2000 years ago as they are now. Other, longer proofs, however, illuminate other aspects of the problem, and lead in other directions . . . In fact, there are many interesting proofs that there are infinitely many primes. Let me tell you two more that I like. You can skim over these two proofs if you wish, or skip them altogether. The important thing is to understand Euclid's original proof!

EULER'S PROOF THAT THERE ARE INFINITELY MANY PRIMES

Warning: this is the most difficult proof in this book. Don't get stuck here. In the future I won't give any long proofs, I'll just explain the general idea. But this is a really beautiful piece of mathematics by a wonderful mathematician. It may not be your favorite way to prove that there are infinitely many primes—it may be overkill—but it shows how very far you can get in a few steps with what is essentially just high-school mathematics. But I really want to encourage you to make the effort to understand Euler's proof. I am not a believer in instant gratification. It is well worth it to make a sustained effort to understand this one piece of mathematics; it's like long, slow foreplay rewarded by a final orgasm of understanding!

First let's sum what's called an infinite geometric series:

$$1 + r + r^2 + r^3 + \ldots = \frac{1}{(1 - r)}$$

This works as long as the absolute value (the unsigned value) of r is less than 1. *Proof:*

Let S_n stand for $1 + r + r^2 + r^3 + \ldots + r^n$.

$$S_n - (r \times S_n) = 1 - r^{n+1}$$

$$\text{Therefore } (1 - r) \times S_n = 1 - r^{n+1}$$

$$\text{and } S_n = \frac{(1 - r^{n+1})}{(1 - r)}.$$

So S_n tends to $\frac{1}{(1 - r)}$ as n goes to infinity,

because if $-1 < r < 1$, then r^{n+1} goes to zero as n goes to infinity (gets bigger and bigger).

So we've just summed an infinite series! Let's check if the result makes any sense. Well, let's look at the special case $r = 1/2$. Then our result says that

$$1 + \frac{1}{2} + \frac{1}{4} + \frac{1}{8} + \frac{1}{16} + \frac{1}{32} + \frac{1}{64} + \ldots + \frac{1}{2^n} + \ldots$$

$$= \frac{1}{(1 - r)} = \frac{1}{\left(\frac{1}{2}\right)} = 2,$$

which is correct. What happens if r is exactly zero? Then the infinite series becomes $1 + 0 + 0 + 0 \ldots = 1$, which is $1/(1 - r)$ with $r = 0$. And if r is just a tiny bit greater than 0, then the sum will be just a tiny bit greater than 1. And what if r is just a tiny little bit less than 1? Then the infinite series starts off looking like $1 + 1 + 1 + 1 \ldots$ and the sum will be very big, which also agrees with our formula: $1/(1 - r)$ is also very large if r is just a tiny bit less than 1. And for $r = 1$ everything falls apart, and coincidentally both the infinite series and our expression $1/(1 - 1) = 1/0$ for the sum give infinity. So that should give us some confidence in the result.

Now we'll sum the so-called "harmonic series" of all reciprocals of the positive integers.

$$1 + \frac{1}{2} + \frac{1}{3} + \frac{1}{4} + \frac{1}{5} \ldots = \infty$$

In other words, it diverges to infinity, it becomes arbitrarily large if you sum enough terms of the harmonic series.

Proof: compare the harmonic series with $1 + 1/2 + 1/4 + 1/4 + 1/8 + 1/8 + 1/8 + 1/8 + \ldots$ Each term of the harmonic series is greater than or equal to the corresponding term in this new series, which is obviously equal to $1 + 1/2 + 1/2 + 1/2 + \ldots$ and therefore diverges to infinity.

But the harmonic series is less than or equal to the product of

$$\frac{1}{\left(1 - \frac{1}{p}\right)} = 1 + \frac{1}{p} + \frac{1}{p^2} + \frac{1}{p^3} + \ldots$$

taken over all the primes p. (This is the sum of a geometric series with ratio $r = 1/p$.) Why? Because the reciprocal of every number can be expressed as a product

$$\frac{1}{(p^\alpha \times q^\beta \times r^\gamma \times \ldots)}$$

of the reciprocal of prime powers p^α, q^β, r^γ ...

In other words,

$$1 + \frac{1}{2} + \frac{1}{3} + \frac{1}{4} + \frac{1}{5} + \frac{1}{6} + \frac{1}{7} + \frac{1}{8} + \frac{1}{9} + \frac{1}{10} + \ldots \leq$$

$$\left(1+\frac{1}{2}+\frac{1}{4}+\frac{1}{8}+\frac{1}{16}+\ldots\right)\times\left(1+\frac{1}{3}+\frac{1}{9}+\frac{1}{27}+\frac{1}{81}+\ldots\right)$$
$$\times\left(1+\frac{1}{5}+\frac{1}{25}+\frac{1}{125}+\frac{1}{625}+\ldots\right)\times\ldots$$

Since the left-hand side of this inequality diverges to infinity, the right-hand side must too, so there must be infinitely many primes!

This leads to Euler's product formula for the Riemann zeta function, as I'll now explain.

Actually, prime factorization is unique. (Exercise for budding mathematicians: Can you prove this by the method of infinite descent? Assume that N is the smallest positive integer that has two different prime factorizations, and show that a smaller positive integer must also have this property. But in fact 1, 2, 3, 4, 5 all have unique factorization, and if we keep going down, we must eventually get down to there! Contradiction!)[3]

So, a special case of what is known as Euler's product formula, $1 + 1/2 + 1/3 + 1/4 + 1/5 \ldots =$ (not $\leq$) product of $1/(1 - 1/p)$ over all primes p. This is a special case of the following more general formula:

$$\zeta(s) = 1 + \frac{1}{2^s} + \frac{1}{3^s} + \frac{1}{4^s} + \frac{1}{5^s} \ldots =$$
$$\text{product over all primes } p \text{ of } \frac{1}{\left(1 - \frac{1}{p^s}\right)},$$

which gives us two different expressions for $\zeta(s)$, which is Riemann's famous zeta function. Above we have been considering $\zeta(1)$, which diverges to infinity. The modern study of the statistical distribution of the primes depends on delicate properties of Riemann's zeta function $\zeta(s)$ for complex arguments $s = a + b\sqrt{-1}$, which is a complicated business that I'm not going to discuss, and is where the famous Riemann hypothesis arises.

[3]For a solution see Courant and Robbins, *What Is Mathematics?*

Let me just tell you how far I got playing with this myself when I was a teenager. You can do a fair amount using comparatively elementary methods and the fact that the sum of the reciprocals of the primes diverges:

$$\frac{1}{2} + \frac{1}{3} + \frac{1}{5} + \frac{1}{7} + \frac{1}{11} + \frac{1}{13} + \frac{1}{17} + \frac{1}{23} + \ldots = \infty$$

This was established by Euler, and it shows that the primes cannot be too sparse, or this infinite series would converge to a finite sum instead of diverging to infinity.

MY COMPLEXITY-BASED PROOF THAT THERE ARE INFINITELY MANY PRIMES

Because if there were only finitely many different primes, expressing a number N via a prime factorization

$$N = 2^e \times 3^f \times 5^g \times \ldots$$

would be too concise! This is too compressed a form to give each number N. There are too many N, and not enough expressions that concise to name them all!

In other words, most N cannot be defined that simply, they are too complex for that. Of course, **some** numbers can be expressed extremely concisely: For example, 2^{99999} is a very small expression for a very large number. And $2^{2^{99999}}$ is an even more dramatic example. But these are **exceptions,** these are atypical.

In general, N requires order of log N characters, but a prime factorization

$$N = 2^e \times 3^f \times 5^g \times \ldots$$

with a fixed number of primes would only require order of log log N characters, and this isn't enough characters to give that many, the required number, of different N!

If you think of $2^e \times 3^f \times 5^g \times \ldots$ as a computer program for generating N, and if there were only finitely many primes, these programs would be too small; they would enable you to compress all N enormously, which is impossible, because in general the best way to specify N via a computer program is to just give it explicitly as a constant in a program with **no** calculation at all!

For those of you who do not know what the "log" function is, you can think of it as "the number of digits needed to write the number N"; that makes it sound less technical. It grows by 1 each time that N is multiplied by ten. log log N grows even more slowly. It increases by 1 each time that log N is multiplied by ten.

Needless to say, I'll explain these ideas much better later. The size of computer programs is one of the major themes of this book. This is just our first taste of this new spice!

DISCUSSION OF THESE THREE PROOFS

Okay, I've just sketched three different very different proofs that there are infinitely many prime numbers. One that's 2000 years old, one that's about 200 years old, and one that's about 20 years old! Notice how very, very different these proofs are!

So right away, I believe that this completely explodes the myth, dear to both Bourbaki and Paul Erdös, that there is only **one** perfect proof for each mathematical fact, just one, the most elegant one. Erdös used to refer to "the book," God's book with the perfect proof of each theorem. His highest praise was, "that's a proof from the book!" As for Bourbaki, that enterprising group of French mathematicians that liked to attribute the output of their collective efforts to the fictitious "Nicolas Bourbaki," they would have endless fights and revisions of their monographs until everything was absolutely perfect. Only perfection was acceptable, nothing less than that!

In my opinion this is a totalitarian doctrine. Mathematical truth is not totally objective. If a mathematical statement is false, there will be no proofs, but if it is true, there will be an endless variety of proofs, not just one! Proofs are not impersonal, they express the personality of their creator/discoverer just as much as literary efforts do. If something important is true, there will be **many** reasons that it is true, many proofs of that fact. Math is the music of reason, and some proofs sound like jazz, others sound like a fugue. Which is better, the jazz or the fugues? Neither: it's all a matter of taste; some people prefer jazz, some prefer fugues, and there's no arguing over individual tastes. In fact this diversity is a good thing: if we all loved the same woman it would be a disaster!

And each proof will emphasize different aspects of the problem, each proof will lead in different directions. Each one will have different corollaries, different generalizations . . . Mathematical facts are not isolated, they are woven into a vast spider's web of interconnections.

As I said, each proof will illuminate a different aspect of the problem. Nothing is ever absolutely black or white; things are always very complicated. Trivial questions may have a simple answer: 2 + 2 is definitely not 5. But if you are asking a **real** question, the answers are more likely to be: "on the one hand this and that, on the other hand so and so," even in the world of pure mathematics, not to mention the real world, which is much, much messier than the imaginary mental mindscape of pure mathematics.

Another thing about the primes and elementary number theory is how close you always are to the frontiers of knowledge. Yes, sometimes you are able to prove something nice, like our three proofs that the list of primes is endless. But those are the good questions, and they are in a minority! Most questions that you ask are extremely difficult or impossible to answer, and even if you can answer them, the answers are extremely complicated and lead you nowhere. In a way, math isn't the art of answering mathematical questions, it is the art of asking the right questions, the questions that give you insight, the ones that lead you in interesting directions, the ones that connect with lots of other interesting questions—the ones with beautiful answers!

And the map of our mathematical knowledge resembles a highway running through the desert or a dangerous jungle; if you stray off the road, you'll be hopelessly lost and die! In other words, the current map of mathematics reflects what our tools are currently able to handle, not what is really out there. Mathematicians don't like to talk about what they don't know, they like to talk about the questions that current technique, current mathematical technology, is capable of handling. Ph.D. students who are too ambitious never finish their thesis and disappear from the profession, unfortunately. And you may think that mathematical reality is objective, that it's not a matter of opinion. Supposedly it is clear whether a proof is correct or not. Up to a point! But whether a piece of math is correct isn't enough, the real question is whether it's "interesting," and that is absolutely and totally a matter of opinion, and one that depends on the current mathematical fashions. So fields become popular, and then they become unpopular, and then they disappear and are forgotten! Not always, but sometimes. Only really important mathematical ideas survive.

Changing direction a bit, let me say why it's good to have many different proofs of an important mathematical result. Each, as I said before, illuminates a different aspect of the problem and reveals different connections and leads you in different directions. But it is also a fact, as was said by the mathematician George Pólya in his lovely little book *How to Solve It* (which I read as a child), that it's better to stand on two legs than on one. If a result is important, you badly want to find different ways to see that fact; that's much safer. If you have only one proof, and it contains an error, then you're left with nothing. Having several proofs is not only safer, it also gives you more insight, and it gives you more understanding. After all, the real goal of mathematics is to obtain insight, not just proofs. A long, complicated proof that gives you no insight is not only psychologically unsatisfying, it's fragile, it may easily turn out to be flawed. And I prefer proofs with ideas, not proofs with lots of computation.

MY LOVE/HATE RELATIONSHIP WITH GÖDEL'S PROOF

> "The theory of numbers, more than any other branch of mathematics, began by being an experimental science. Its most famous theorems have all been conjectured, sometimes a hundred years or more before they were proved; and they have been suggested by the evidence of a mass of computations."
>
> —*G. H. Hardy,* quoted in Dantzig, op. cit.

So number theory is an experimental science as well as a mathematical theory. But theory lags far, far behind experiment! And one wonders, will it ever catch up? Will the primes refuse to be tamed? Will the mystery remain? I was fascinated reading about all of this as a child.

And then one day I discovered that a little book had just been published. It was by Nagel and Newman, and it was called *Gödel's Proof.* This was in 1958, and the book was an expanded version of an article that I'd also seen, and that was published by the two of them in *Scientific American* in 1956. It was love at first sight! Mad love, crazy love, obsessive love, what the French call *amour à la folie.* Here in fact was a possible explanation for the difficulties that mathematicians were experiencing with the primes: Gödel's incompleteness theorem, which asserts that any finite system of mathematical axioms, any mathematical theory, is **incomplete.** More precisely, he showed that there will always be arithmetic assertions, assertions about the positive integers and addition and multiplication, what are called number-theoretic assertions, that are true but unprovable!

I carried this book around with me constantly, absolutely and totally fascinated, mesmerized by the whole idea. There was only one small, tiny little problem, fortunately, which was that for the life of me I couldn't understand Gödel's proof of this wonderful metamathematical result. It's called that because it's not a mathematical result, it's a theorem **about** mathematics itself, about the limitations of mathematical methods. It's not a result within any field of mathematics, it stands outside looking down at mathematics, which is itself a field called metamathematics!

I wasn't an idiot, so why couldn't I understand Gödel's proof? Well, I could follow it step by step, but it was like trying to mix oil and water. My mind kept resisting. In other words, I didn't lack the necessary intelligence, I just plain didn't like Gödel's proof of his fabulous result. His original proof seemed too complicated, too fragile! It didn't seem to get to the heart of the matter, because it was far from clear how prevalent incompleteness might in fact be.

And this is where my own career as a mathematician takes off. I loved reading about number theory, I loved computer programming (for example, programs for calculating primes), I disliked Gödel's proof, but I loved Turing's alternative approach to incompleteness using the idea of the computer. I felt very comfortable with computers. I thought they were a great toy, and I loved writing, debugging and running computer programs—FORTRAN programs, machine language programs—I thought that this was great mental entertainment! And at 15 I had the idea—anticipated by Leibniz in 1686—of looking at the size of computer programs and of defining a random string of bits to be one for which there is no program for calculating it that is substantially smaller than it is.

The idea was define a kind of logical, mathematical or structural randomness, as opposed to the kind of physical randomness that Einstein and Infeld's delightful book *The Evolution of Physics* emphasized to be an essential characteristic of quantum physics, the physics of the microcosm. That's another book suitable for children and teenagers that I can highly recommend, and that I also worshiped at that stage of my life.

Let me explain what happened better; I'll now reveal to you one of the secrets of mathematical creation! I loved incompleteness, but not Gödel's proof. Why? Because of the lack of balance between the ends and the means, between the theorem and its proof. Such a deep and important—philosophically important—mathematical result deserved a deep proof that would give deep insight into the "why" of incompleteness, instead of a clever proof that only permitted you to have a superficial understanding of what was going on. That was my feeling, totally on intuitive grounds, pure instinct, pure intuition, my subconscious, gut-level, emotional reaction to Gödel's proof.

And so I set to work to make it happen! This was a totally subjective act of creation, because I **forced** it to happen. How? Well, by changing the rules of the game, by reformulating the problem, by redefining the context in which incompleteness was discussed in such a way that there would **be** a deep reason for incompleteness, in such a way that a deeper reason for incompleteness could emerge! You see, within the context that Gödel worked, he had done the best that was possible. If you were to keep the setup exactly the same as the one he had dealt with, there **was no** deeper reason for incompleteness. And so I proceeded to change the question until I could get out a deep reason for incompleteness. My instinct was that the original context in which the problem of incompleteness was formulated had to be changed to one that permitted such deeper understanding—that it was the wrong context if this wasn't possible!

Now you see why I say that the mathematician is a creator as much as a discoverer, and why I say that mathematical creation is a totally personal act.

On the other hand, I couldn't have succeeded if my intuitions that there was a deeper reason were incorrect. Mathematical truth is malleable, but only up to a point!

Another way to put it is that I wanted to eliminate the superficial details that I thought obscured these deeper truths, and I proceeded to change the formulation of the problem in order to make this happen. So you might say that this is an act of pure invention, that I created a deeper reason for incompleteness because I so badly wanted there to be one. That's true up to a point. Another way to put it is that my intuition whispered to me because the idea **wanted** to be found, because it was a more natural, a less forced way to perceive the question of incompleteness. So, from that point of view, I wasn't **making** anything happen, on the contrary! I simply was acutely sensitive to the vague half-formulated but more natural view of incompleteness that was partially obscured by Gödel's original formulation.

I think that both views of this particular act of creation are correct: On the one hand, there was a *masculine* component, in making something happen by ignoring the community consensus of how to think about the problem. On the other hand, there was a *feminine* component, in allowing my hypersensitive intuition to sense a delicate new truth that no one else was receptive to, that no one else was listening for.

The purpose of this book is to explain what I created/discovered to you. It took many years of work, culminating with the halting probability Ω—sometimes called Chaitin's number—that's the discovery that I'm most proud of. It will take several chapters for me to explain all of this well, because I'll have to build the appropriate intellectual framework for thinking about incompleteness and my Ω number.

The first step is for me to explain to you the starting point for my own work, which most definitely was not Gödel's proof of 1931, but was instead Turing's alternative 1936 approach to incompleteness, in which the idea of computation plays a fundamental role. And it's thanks to Turing that the idea of computation, that the idea of the computer, became a new force in mathematical thinking. So let me tell you how this happened, and then I'll illustrate the power of this new idea by showing you the way in which it solved an outstanding open question called Hilbert's 10th problem, that was formulated by Hilbert as one of a list of 23 challenge problems in 1900. That'll keep us busy for the rest of this chapter.

In the next chapter I'll get back to my fundamental new idea, to what I added to Turing, which is my definition of randomness and complexity that Leibniz—who happens to be the inventor of the infinitesimal calculus—clearly and definitely anticipated in 1686.

HILBERT, TURING & POST ON FORMAL AXIOMATIC SYSTEMS & INCOMPLETENESS

The first step in the direction of being able to use math to study the power of math was taken by David Hilbert about a century ago. So I view him as the creator of metamathematics. It was his idea that in order to be able to study what mathematics can achieve, we first have to specify completely the rules of the game. It was his idea to create a formal axiomatic system or FAS for all of mathematics, one that would eliminate all the vagueness in mathematical argumentation, one that would eliminate any doubt whether a mathematical proof is correct or not.

How is this done? What's in a formal axiomatic system? Well, the general idea is that it's like Euclid's *Elements,* except that you have to be much, much more fussy about all the details!

The first step is to create a completely formal artificial language for doing mathematics. You specify the alphabet of symbols that you're using, the grammar, the axioms, the rules of inference, and a proof-checking algorithm:

Hilbert Formal Axiomatic System

Alphabet
Grammar
Axioms
Rules of Inference
Proof-Checking Algorithm

Mathematical proofs must be formulated in this language with **nothing** missing, with every tiny step of the reasoning in place. You start with the axioms, and then you apply the rules of inference one by one, and you deduce all the theorems! This was supposed to be like a computer programming language: So precise that a machine can interpret it, so precise that a machine can understand it, so precise that a machine can check it. No ambiguity, none at all! No pronouns. No spelling mistakes! Perfect grammar!

And part of this package is that a finite set of mathematical axioms or postulates are explicitly given, plus you use symbolic logic to deduce all the possible consequences of the axioms. The axioms are the starting point for any mathematical theory; they are taken as self-evident, without need for proof. The consequences of these axioms, and the consequences of the consequences, and the consequences of that, and so forth and so on *ad infinitum,* are called the "theorems" of the FAS.[4]

And a key element of a FAS is that there is a proof-checking algorithm, a mechanical procedure for checking if a proof is correct or not. In other words, there is a computer program that can decide whether or not a proof follows all the rules. So if you are using a FAS to do mathematics, then you do not need to have human referees check whether a mathematics paper is correct before you publish it. You just run the computer program, and it tells you whether or not there's a mistake!

Up to a point this doesn't seem too much to demand: It's just the idea that math can achieve perfect rigor, that mathematical truth is black or white, that math provides absolute certainty. We will see!

Math = Absolute Certainty???

The first step in this drama, in the decline and fall of mathematical certainty, was an idea that I learned from Alan Turing's famous 1936 paper introducing the computer—as a mathematical idea, not with actual hardware! And the most interesting thing about this paper is Turing's famous halting problem, which shows that there are things that no computer can ever calculate, no matter how cleverly you program it, no matter how patient you are in waiting for an answer. In fact Turing finds two such things: the halting problem; and uncomputable real numbers, which we'll discuss in Chapter Five.

Turing, 1936: *Halting Problem*

Here I'll just talk about the halting problem. What's the halting problem? It's the question of whether or not a computer program,

[4]These ingredients are already in Euclid, except that some of his axioms are tacit, some steps in his proofs are skipped, and he uses Greek instead of symbolic logic, in other words, a human language, not a machine language.

one that's entirely self-contained, one that does no input/output, will ever stop. If the program needs any numbers, they have to be given in the program itself, not read in from the outside world.

So the program grinds away step by step, and either goes on forever, or else it eventually stops, and the problem is to decide which is the case in a finite amount of time, without waiting forever for it to stop.

And Turing was able to show the extremely fundamental result that there is no way to decide in advance whether or not a computer program will ever halt, not in a finite amount of time. If it does halt, you can eventually discover that. The problem is to decide when to give up and decide that it will never halt. But there is no way to do that.

Just like I'm not polluting your mind with Gödel's proof, I won't say a word about how Turing showed that the halting problem cannot be solved, that there is no algorithm for deciding if a computer program will never halt.

I'll give you my own proof later, in Chapter Six. I'll show that you can't prove that a program is "elegant," by which I mean that it's the smallest possible program for producing the output that it does. From the fact that you can't establish elegance, I'll deduce the immediate consequence that the halting problem must be unsolvable. Not to worry, it'll all be explained in due course.

But what I will explain right now is how Turing derived incompleteness from the halting problem.

Turing: *Halting Problem Implies Incompleteness!*

The idea is very simple. Let's assume that we have a FAS that can always **prove** whether or not individual programs halt. Then you just run through all possible proofs in size order until you find a proof that the particular program that you're interested in never halts, or you find a proof that it does in fact halt. You work your way systematically through the tree of all possible proofs, starting from the axioms. Or you can write down one by one all the possible strings of characters in your alphabet, in size order, and apply the proof-checking algorithm to them to filter out the invalid proofs and determine all the valid theorems. In either case it's slow work, very slow, but what do I care, I'm a theoretician! This isn't supposed to be a practical approach; it's more like what physicists call a thought experiment, *Gedankenexperiment* in the original German. I'm trying to prove a theorem, I'm trying to show that a FAS must have certain limitations, I'm not trying to do anything practical.

So, Turing points out, if we have a FAS that can always prove whether or not individual programs halt, and if the FAS is "sound" which means that all the theorems are true, then we can generate the theorems of the FAS one by one and use this to decide if any particular program that we are interested in will ever halt! But that's impossible, it cannot be, as Turing proved in his 1936 paper and I'll prove in a totally different manner in Chapter Six.

So the FAS must be incomplete. In other words, there must be programs for which there is no proof that the program halts and there is also no proof that it will never halt. There is no way to put all the truth and only the truth about the halting problem into a FAS!

Turing's marvelous idea was to introduce the notion of computability, of distinguishing things that cannot be calculated from those that can, and then deduce incompleteness from uncomputability.

Turing: *Uncomputability Implies Incompleteness!*

Uncomputability is a deeper reason for incompleteness. This immediately makes incompleteness seem much more natural, because, as we shall see in Chapter Five, a lot of things are uncomputable, they're everywhere, they're very easy to find.

(Actually, what Turing has shown is more general than that, it's that

Soundness + Completeness implies you can systematically settle *anything* you can ask your FAS!

In fact, in some limited domains you **can** do that: Tarski did it for a big chunk of Euclidean geometry.)

My own approach to incompleteness is similar to Turing's in that I deduce incompleteness from something deeper, but in my case it'll be from randomness, not from uncomputability, as we shall see. And random things are also everywhere, they're the rule, not the exception, as I'll explain in Chapter Five.

My Approach: *Randomness Implies Incompleteness!*

Anyway, the next major step on this path was taken in 1944 by Emil Post, who happens to have been a professor at the City College of the City University of New York, which is where I was a student when I wrote my first major paper on randomness. They were very impressed by this paper at City College, and they were nice enough to give me a gold medal, the Belden Mathematical Prize, and later the Nehemiah Gitelson award for "the search for truth." It says that on the Gitelson medal in English and Hebrew, which is how I happen to know that Truth and Law are both Torah in Hebrew; that's crucial in order to be able to understand the Kafka parable at the beginning of this book. And as the extremely kind head of the math department, Professor Abraham Schwartz, was handing me the Belden Prize gold medal, he pointed to the photos of former professors hanging on the wall of his office, one of whom was Emil Post.

Post had the insight and depth to see the key idea in Turing's proof that uncomputability implies incompleteness. He extracted that idea from Turing's proof. It was a jewel that Turing himself did not sufficiently appreciate. Post sensed that the essence of a FAS is that there is an algorithm for generating all the theorems, one that's very slow and that never stops, but that eventually gets to each and every one of them. This amazing algorithm generates the theorems in size order, but not in order of the size of the statement of each theorem, in order of the size of each proof.

Hilbert/Turing/Post Formal Axiomatic System

Machine for generating all the theorems
one by one, in some arbitrary order.

Of course, as I believe Émile Borel pointed out, it would be better to generate just the interesting theorems, not all the theorems; most of them are totally uninteresting! But no one knows how to do that. In fact, it's not even clear what an "interesting" theorem is.[5]

Anyway, it was Post who put his finger on the essential idea, which is the notion of a set of objects that can be generated one by one by a machine, in some order, any order. In other words, there is an algorithm, a computer program, for doing this. And that's the essential content of the notion of a FAS, that's the toy model that I'll use to study the limits of the formal axiomatic method. I don't care about the details, all that matters to me is that there's an algorithm for generating all the theorems, that's the key thing.

And since this is such an important notion, it would be nice to give it a name! It used to be called an r.e. or "recursively enumerable" set. The experts seem to have switched to calling it a "computably enumerable" or c.e. set. I'm tempted to call it something that Post said in one of his papers, which is that it's a "generated set," one that can be generated by an algorithm, item by item, one by one, in some order or other, slowly but surely. But I'll resist the temptation!

[5]Wolfram has some thoughts about what makes theorems interesting in his book. As usual, he studies a large number of examples and extracts the interesting features. That's his general *modus operandi.*

FAS = c.e. set of mathematical assertions

So what's the bottom line? Well, it's this: Hilbert's goal of **one** FAS for all of math was an impossible goal, because you can't put all of mathematical truth into just one FAS. Math can't be static, it's got to be dynamic, it's got to evolve. You've got to keep extending your FAS, adding new principles, new ideas, new axioms, just as if you were a physicist, without proving them, because they work! Well, not exactly the way things are done in physics, but more in that spirit. And this means that the idea of absolute certainty in mathematics becomes untenable. Math and physics may be different, but they're not that different, not as different as people might think. Neither of them gives you absolute certainty!

We'll discuss all of this at greater length in the concluding chapter, Chapter Seven. Here just let me add that I personally view metamathematics as a *reductio ad absurdum* of Hilbert's idea of a formal axiomatic system. It's an important idea **precisely because** it can be shot down! And in the conflict between Hilbert and Poincaré over formalism versus intuition, I am now definitely on the side of intuition.

And as you have no doubt noticed, here we've been doing "philosophical" math. There are no long proofs. All that counts are the ideas. You just take care of the ideas, and the proofs will take care of themselves! That's the kind of math I like to do! It's also beautiful mathematics, just as beautiful as proving that there are infinitely many primes. But it's a different kind of beauty!

Armed with these powerful new ideas, let's go back to number theory. Let's see if we can figure out why number theory is so hard.

We'll use this notion of a computably enumerable set. We'll use it to analyze Hilbert's 10th problem, an important problem in number theory.

More precisely, I'll explain what Hilbert's 10th problem is, and then I'll tell you how Yuri Matiyasevich, Martin Davis, Hilary Putnam and Julia Robinson managed to show that it **can't** be solved.

Davis, Putnam and Robinson did much of the preparatory work, and then the final steps were taken by Matiyasevich. Some important additional results were then obtained much more simply by Matiyasevich together with James Jones. Matiyasevich and Jones built on a curious and unappreciated piece of work by Édouard Lucas, a remarkable and extremely talented and unconventional French mathematician of the late 1800s who never let fashion stand in his way. Lucas also invented an extremely fast algorithm for deciding if a Mersenne number $2^n - 1$ is a prime or not, which is how the largest primes that we currently know were all discovered.

By the way, Martin Davis studied with Emil Post at City College. It's a small world!

HILBERT'S 10TH PROBLEM & DIOPHANTINE EQUATIONS AS COMPUTERS

What is a diophantine equation? Well, it's an algebraic equation in which everything, the constants as well as the unknowns, has got to be an integer. And what is Hilbert's 10th problem? It's Hilbert's challenge to the world to discover an algorithm for determining whether a diophantine equation can be solved. In other words, an algorithm to decide if there is a bunch of whole numbers that you can plug into the unknowns and satisfy the equation.

Note that if there **is** a solution, it can eventually be found by systematically plugging into the unknowns all the possible whole-number values, starting with small numbers and gradually working your way up. The problem is to decide when to give up. Doesn't that sound familiar? Yes, that's exactly the same as what happens with Turing's halting problem!

Diophantus was, like Euclid, a Greek scholar in Alexandria, where the famous library was. And he wrote a book about diophantine equations that inspired Fermat, who read a Latin translation of the Greek original. Fermat's famous "last theorem," recently proved by Andrew Wiles, was a marginal annotation in Fermat's copy of Diophantus.

More precisely, we'll work only with unsigned integers, namely the positive integers and zero: 0, 1, 2, 3, …

For example, $3 \times x \times y = 12$ has the solution $x = y = 2$, and $x^2 = 7$ has no whole-number solution.

And we'll allow an equation to be assembled from constants and unknowns using only additions and multiplications, which is called an (ordinary) diophantine equation, or using additions, multiplications and exponentiations, which is called an exponential diophantine equation. The idea is to avoid negative integers and minus signs and subtraction by actually using both sides of the equation. So you **cannot** collect everything on the left-hand side and reduce the right-hand side to zero.

In an ordinary diophantine equation the exponents will always be constants. In an exponential diophantine equation, exponents may also contain unknowns. Fermat-Wiles states that the exponential diophantine equation

$$x^n + y^n = z^n$$

has no solutions with x, y and z greater than 0 and n greater than 2. It only took a few hundred years to find a proof! Too bad that Fermat only had enough room in the margin of Diophantus to state his result, but not enough room to say **how** he did it. And all of Fermat's personal papers have disappeared, so there is nowhere left to search for clues.

So that's just how very bad an exponential diophantine problem can be! Hilbert wasn't that ambitious, he was only asking for a way to decide if **ordinary** diophantine equations have solutions. That's bad enough!

To illustrate these ideas, let's suppose we have two equations, $u = v$ and $x = y$, and we want to combine them into a single equation that has precisely the same solutions. Well, if we could use subtraction, the trick would be to combine them into

$$(u - v)^2 + (x - y)^2 = 0.$$

The minus signs can be avoided by converting this into

$$u^2 - (2 \times u \times v) + v^2 + x^2 - (2 \times x \times y) + y^2 = 0$$

and then to

$$u^2 + v^2 + x^2 + y^2 = (2 \times u \times v) + (2 \times x \times y).$$

That does the trick!

Okay, now we're ready to state the amazing Matiyasevich/Davis/Putnam/Robinson solution of Hilbert's 10th problem. Diophantine equations come, as we've seen, from classical Alexandria. But the solution to Hilbert's problem is amazingly modern, and could not even have been imagined by Hilbert, since the relevant concepts did not yet exist.

Here is how you show that there is no solution, that there is no algorithm for determining whether or not a diophantine equation can be solved, that there will never be one.

It turns out that there is a diophantine equation that is a computer. It's actually called a universal diophantine equation, because it's like a universal Turing machine, math-speak for a general-purpose computer, one that can be programmed to run any algorithm, not just perform special-purpose calculations like some pocket calculators do. What do I mean? Well, here's the equation:

Diophantine Equation Computer:

$$L(k, n, x, y, z, \ldots) = R(k, n, x, y, z, \ldots)$$

Program k
Output n
Time $x, y, z, \ldots$

(The real thing is too big to write down!) There's a left-hand side L, a right-hand side R, a lot of unknowns $x, y, z, \ldots$ and two special symbols, k, called the parameter of the equation, and n, which is the unknown that we really care about. **k is the program for the computer, n is the output that it produces, and $x, y, z, \ldots$ are a multi-dimensional time variable!** In other words, you put into the equation $L(k, n) = R(k, n)$ a specific value for k which is the computer program that you are going to run. Then you focus on the values of n for which there are solutions to the equation, that is, for which you can find a time $x, y, z, \ldots$ at which the program k outputs n. That's it! That's all there is to it!

So, amazingly enough, this **one** diophantine equation can perform **any** calculation. In particular, you can put in a k that's a program for calculating all the primes, or you can put in a k that's a program for computably enumerating all the theorems of a particular FAS. The theorems will come out as numbers, not character strings, but you can just convert the number to binary, omit the leftmost 1 bit, divide it up into groups of 8 bits, look up each ASCII character code, and presto chango, you have the theorem!

And this diophantine equation means we're in trouble. It means that there **cannot** be a way to decide if a diophantine equation has any solutions. Because if we could do that, we could decide if a computer program gives any output, and if we could do that, we could solve Turing's halting problem.

Halting Problem Unsolvable Implies Hilbert's 10th Problem Is Unsolvable!

Why does being able to decide if a computer program has any output enable us to decide whether or not a program halts? Well, any program that either halts or doesn't and produces no output can be converted into one that produces a message "I'm about to halt!" (in binary, as an integer) just before it does. If it's written in a high-level language, that should be easy to do. If it's written in a machine language, you just run the program interpretively, not directly, and catch it just before it halts. So if you can decide whether the converted program produces any output, you can decide whether the original program halts.

That's all there is to it: diophantine equations are actually computers! Amazing, isn't it! What wouldn't I give to be able to explain this to Diophantus and to Hilbert! I'll bet they would understand right away! (This book is how I'd do it.)

And this *proves* that number theory is hard! This proves that uncomputability and incompleteness are lurking right at the core, in two-thousand-year-old diophantine problems! Later I'll show that randomness is also hiding there . . .

And how do you actually construct this monster diophantine equation computer? Well, it's a big job, rather like designing actual computer hardware. In fact, I once was involved in helping out on the design of an IBM computer. (I also worked on the operating system and on a compiler, which was a lot of fun.) And the job of constructing this diophantine equation reminds me of what's called the logical design of a CPU. It's sort of an algebraic version of the design of a CPU.

The original proof by Matiyasevich/Davis/Putnam/Robinson is complicated, and designs an ordinary diophantine equation computer. The subsequent Matiyasevich/Jones design for an exponential diophantine equation computer is much, much simpler—fortunately!—so I was able to actually program it out and exhibit the equation, which nobody else had ever bothered to do. I'll tell you about that in the next section, the section on LISP.

And in designing this computer, you don't work with individual bits, you work with strings of bits represented in the form of integers, so you can process a lot of bits at a time. And the key step was provided by Lucas a century ago and is a way to compare two bit strings that are the same size and ensure that every time that a bit is on in the first string, the corresponding bit is also on in the second string. Matiyasevich and Jones show that that's enough, that it can be expressed via diophantine equations, and that it's much more useful to be able to do it than you might think. In fact, that's all the number theory that Matiyasevich and Jones need to build their computer; the rest of the job is, so to speak, computer engineering, not number theory. From that point on, it's basically just a lot of high-school math!

So what's this great idea contributed posthumously by Lucas? Well, it involves "binomial coefficients." Here are some. They're what you get when you expand powers of $(1 + x)$:

Binomial Coefficients:

$$(1 + x)^0 = 1$$
$$(1 + x)^1 = 1 + x$$
$$(1 + x)^2 = 1 + 2x + x^2$$
$$(1 + x)^3 = 1 + 3x + 3x^2 + x^3$$
$$(1 + x)^4 = 1 + 4x + 6x^2 + 4x^3 + x^4$$
$$(1 + x)^5 = 1 + 5x + 10x^2 + 10x^3 + 5x^4 + x^5$$

Let's consider the coefficient of x^k in the expansion of $(1 + x)^n$. Lucas's amazing result is that this binomial coefficient is odd if and only if each time that a bit is on in the number k, the corresponding bit is also on in the number n!

Let's check some examples. Well, if $k = 0$, then every binomial coefficient is 1 which is odd, no matter what n is, which is fine, because every bit in k is off. And if $k = n$, then the binomial coefficient is also 1, which is fine. How about $n = 5 =$ "101" in binary, and $k = 1 =$ "001." Then the binomial coefficient is 5 which is odd, which is correct. What if we change k to $2 =$ "010." Aha! Then each bit that's on in k is **not** on in n, and the binomial coefficient is 10, which is **even**! Perfect!

Please note that if the two bit strings aren't the same size, then you have to add 0's **on the left** to the smaller one in order to make them the same size. Then you can look at the corresponding bits.

[*Exercise for budding mathematicians:* Can you check more values? By hand? On a computer? Can you prove that Lucas was right?! It shouldn't be too difficult to convince yourself. As usual, it's much more difficult to **discover** a new result, to **imagine** the happy possibility, than to verify that it's correct. Imagination, inspiration and hope are the key! Verification is routine. **Anyone** can do that, any competent professional. I prefer tormented, passionate searchers for Truth! People who have been seized by a demon!—hopefully a **good** demon, but a demon nevertheless!]

And how do Matiyasevich and Jones express (if a bit is on in k it also has to be on in n) as a diophantine equation?

Well, I'll just show you how they did it, and you can see why it works with the help of the hints that I'll give.

The coefficient of x^K in the expansion of $(1 + x)^N$ is odd if and only if there is a **unique** set of seven unsigned whole numbers b, x, y, z, u, v, w such that

$$b = 2^N$$

$$(b+1)^N = xb^{K+1} + yb^K + z$$

$$z + u + 1 = b^K$$
$$y + v + 1 = b$$
$$y = 2w + 1$$

Can you figure out how this works? *Hint:* y is the binomial coefficient that we're interested in, w is used to ensure that y is odd, u guarantees that z is smaller than b^K, v guarantees that y is smaller than b, and when $(b + 1)^N$ is written as a base b numeral its successive digits are precisely the binomial coefficients that you get by expanding $(1 + x)^N$. The basic idea here is that the powers of the number eleven give you the binomial coefficients: $11^2 = 121$, $11^3 = 1331$, $11^4 = 14641$. Then things fall apart because the binomial coefficients get too large to be individual digits. So you have to work in a large base b. The particular b that Matiyasevich and Jones pick happens to be the **sum** of all the binomial coefficients that we are trying to pack together into the digits of a single number. So the base b is larger than all of them, and there are no carries from one digit to the next to mess things up, so everything works!

And then you combine these five individual equations into a **single** equation the way I showed you before. There will be ten squares on one side, and five cross-products on the other side . . .

Now that we've done this, we have the fundamental tool that we need in order to be able to build our computer. The CPU of that computer is going to have a bunch of hardware registers, just like real computers do. But unlike real computers, each register will hold an unsigned integer, which can be an arbitrarily large number of bits. And the unknowns in our master equation will give the contents of these registers. Each digit of one of these unknowns will be the number that that register happened to have at a particular moment, step by step, as the CPU clock ticks away and it executes one instruction per clock cycle. In other words, you have to use the same trick of packing **a list** of numbers into a single number that we used before with binomial coefficients. Here it's done with the time history of CPU register contents, a register contents history that ranges from the start of the computation to the moment that you output an integer.

I'm afraid that we're going to have to stop at this point. After all, constructing a computer is a big job! And I'm personally much more interested in general ideas than in details. You just take care of the general ideas, and the details will take care of themselves! Well, not always: You wouldn't want to buy a computer that was designed that way! But carrying out this construction in detail is a big job . . . Not much fun! Hard work!

Actually, it **is** a lot of fun if you do it yourself. Like making love, it's not much fun to hear how somebody else did it. You have to do it yourself! And you don't learn much by reading someone else's software. It's excruciating to have to do that, you're lost in the maze of someone else's train of thought! But if you yourself write the program and debug it on interesting examples, then it's **your** train of thought and you learn a lot. Software is frozen thought . . .

So we won't finish working out the design in detail. But the basic ideas are simple, and you can find them in the short paper by Matiyasevich and Jones, or in a book I wrote that was published by Cambridge University Press in 1987. It's called *Algorithmic Information Theory,* and that's where I actually exhibit one of these computer equations. It's an equation that runs LISP programs.

What kind of a programming language is LISP? Well, it's a very beautiful language that's highly mathematical in spirit.

LISP, AN ELEGANT PROGRAMMING LANGUAGE

Utopian social theorists and philosophers have often dreamt of a perfect, ideal universal human language that would be unambiguous and would promote human understanding and international cooperation: for example, Esperanto, Leibniz's *characteristica universalis,* Hilbert's formal axiomatic system for all of mathematics. These utopian fantasies have never worked. But they have worked for computers—very, very well indeed!

Unfortunately, as programming languages become increasingly sophisticated, they reflect more and more the complexity of human society and of the immense world of software applications. So they become more and more like giant toolboxes, like garages and attics stuffed with 30 years of belongings! On the contrary, LISP is a programming language with considerable mathematical beauty; it is more like a surgeon's scalpel or a sharp-edged diamond-cutting tool than a two-car garage overflowing with possessions and absolutely no room for a car.

LISP has a few powerful simple basic concepts, and everything else is built up from that, which is how mathematicians like to work; it's what mathematical theories look like. Mathematical theories, the good ones, consist in defining a few new key concepts, and then the fireworks begin: they reveal new vistas, they open the door to entirely new worlds. LISP is like that too; it's more like pure math than most programming languages are. At least it is if you strip away the "useful" parts that have been added on, the accretions that have made LISP into a "practical tool." What's left if you do that is the original

LISP, the conceptual heart of LISP, a core which is a jewel of considerable mathematical beauty, austere intellectual beauty.

So you see, for me this is a real love affair. And how did I fall in love with LISP? In 1970 I was living in Buenos Aires, and I bought a LISP manual in Rio de Janeiro. On page 14 there was a LISP interpreter written in LISP. I didn't understand a word! It looked extremely strange. I programmed the interpreter in FORTRAN and suddenly saw the light. It was devastatingly beautiful! I was hooked! (I still am. I've used LISP in six of my books.) I ran lots of LISP programs on my interpreter. I wrote dozens of LISP interpreters, probably hundreds of LISP interpreters! I kept inventing new LISP dialects, one after another . . .

You see, learning a radically new programming language is a lot of work. It's like learning a foreign language: you need to pick up the vocabulary, you need to acquire the culture, the foreign world-view, that inevitably comes with a language. Each culture has a different way of looking at the world! In the case of a programming language, that's called its "programming paradigm," and to acquire a new programming paradigm you need to read the manual, you need to see a lot of examples, and you need to try writing and running programs yourself. We can't do all that here, no way! So the idea is just to whet your appetite and try to suggest how very different LISP is from normal programming languages.

HOW DOES LISP WORK?

First of all, LISP is a non-numerical programming language. Instead of computing only with numbers, you deal with symbolic expressions, which are called "S-expressions."

Programs and data in LISP are both S-expressions. What's an S-expression? Well, it's like an algebraic expression with full parenthesization. For example, instead of

$$a \times b + c \times d$$

you write

$$((a \times b) + (c \times d))$$

And then you move operators forward, in front of their arguments rather than between their arguments.

$$(+ (\times a\, b)\, (\times c\, d))$$

That's called prefix as opposed to infix notation. Actually in LISP it's written this way

```
(+ (* a b) (* c d))
```

because, as in many other programming languages, * is used for multiplication. You also have minus − and exponentiation ^ in LISP.

Let's get back to S-expressions. In general, an S-expression consists of a nest of parentheses that balance, like this

```
( ( ) (( )) ((( ))) )
```

What can you put inside the parentheses? Words, and unsigned integers, which can both be arbitrarily big. And for doing math with integers, which are exact, not approximate numbers, it's very important to be able to handle very large integers.

Also, a word or unsigned integer can appear all by itself as an S-expression, with **no** parentheses. Then it's referred to as an atom (which means "indivisible" in Greek). If an S-expression isn't an atom, then it's called a list, and it's considered to be a list of elements, with a first element, a second, a third, etc.

```
(1 2 3) and ((x 1) (y 2) (z 3))
```

are both lists with three elements.

And LISP is a functional or expression-based language, not an imperative or statement-based language.

Everything in LISP (that's instructions rather than data) is built up by applying functions to arguments like this:

```
(f x y)
```

This indicates that the function *f* is applied to the arguments *x* and *y*. In pure math that's normally written

$$f(x,y)$$

And **everything** is put into this function applied to arguments form.

For example,

```
(if condition true-value false-value)
```

is how you choose between two values depending on whether a condition is true or not. So "if" is treated as a three-argument function with the strange property that only two of its arguments are evaluated. For instance

```
(if true (+ 1 2) (+ 3 4))
```

gives 3, and

```
(if false (+ 1 2) (+ 3 4))
```

gives 7. Another pseudo-function is the quote function, which doesn't evaluate its only argument. For instance

```
(' (a b c))
```

gives `(a b c)` "as is." In other words, this does **not** mean that the function *a* should be applied to the arguments *b* and *c*. The argument of quote is literal data, not an expression to be evaluated.

Here are two conditions that may be used in an "if."

```
(= x y)
```

gives true if x is equal to y.

```
(atom x)
```

gives true if x is an atom, not a list.

Next let me tell you about "let," which is very important because it enables you to associate values with variables and to define functions.

```
(let x y expression)
```

yields the value of "expression" in which x is defined to be y. You can use "let" either to define a function, or like an assignment statement in an ordinary programming language. However, the function definition or assignment is only temporarily in effect, inside "expression." In other words, the effect of "let" is invariably local.

Here are two examples of how to use "let." Let n be $1 + 2$ in $3 \times n$:

```
(let n (+ 1 2)
   (* 3 n)
)
```

This gives 9. And let f of n be $n \times n$ in f of 10:

```
(let (f n) (* n n)
   (f 10)
)
```

This gives 100.

And in LISP we can take lists apart and then reassemble them. To get the first element of a list, "car" it:

```
(car (' (a b c)))
```

gives a. To get the rest of a list, "cdr" it:

```
(cdr (' (a b c)))
```

gives `(b c)`. And to reassemble the pieces, “cons” it:

```
(cons (' a) (' (b c)))
```

gives `(a b c)`.

And that’s it, that gives you the general idea! LISP is a simple but powerful formalism.

Before showing you two real LISP programs, let me point out that in LISP you don’t speak of programs, they’re called expressions. And you don’t run them or execute them, you evaluate them. And the result of evaluating an expression is merely a value; there is no side-effect. The state of the universe is unchanged.

Of course, mathematical expressions have always behaved like this. But normal programming languages are not at all mathematical. LISP **is** mathematical.

In this chapter we’ve used the factorial function “!” several times. So let’s program that in a normal language, and then in LISP. Then we’ll write a program to take the factorials of an entire list of numbers, not just one.

FACTORIAL IN A NORMAL LANGUAGE

Recall that $N! = 1 \times 2 \times 3 \times \ldots \times (N - 1) \times N$. So $3! = 3 \times 2 \times 1 = 6$, and $4! = 24$, $5! = 120$. This program calculates 5 factorial:

```
Set N equal to 5.
Set K equal to 1.

LOOP: Is N equal to 0? If so, stop. The result is
      in K.
If not, set K equal to K × N.
Then set N equal to N - 1.
Go to LOOP.
```

When the program stops, *K* will contain 120, and *N* will have been reduced to 0.

FACTORIAL IN LISP

The LISP code for calculating factorial of 5 looks rather different:

```
(let (factorial N)
         (if (= N 0)
              1
              (* N (factorial (- N 1)))
         )
 (factorial 5)
)
```

This gives

```
120
```

Here's the definition of the function factorial of *N* in English: If *N* is 0, then factorial of *N* is 1, otherwise factorial of *N* is *N* times factorial of *N* minus 1.

A MORE SOPHISTICATED EXAMPLE: LIST OF FACTORIALS

Here's a second example. Given a list of numbers, it produces the corresponding list of factorials:

```
(let (factorial N)
         (if (= N 0)
              1
              (* N (factorial (- N 1)))
         )
```

```
(let (map f x)
    (if (atom x)
        x
        (cons (f (car x))
              (map f (cdr x))
        )
    )

 (map factorial (' (4 1 3 2 5)))

))
```

This gives

```
(24 1 6 2 120)
```

The "map" function changes $(x\ y\ z\ \ldots)$ into $(f(x)\ f(y)\ f(z)\ \ldots)$. Here's the definition of "map" f of x in English: The function "map" has two arguments, a function f and a list x. If the list x is empty, then the result is the empty list. Otherwise it's the list beginning with (f of the first element of x) followed by "map" f of the rest of x.

[For more on my conception of LISP, see my books *The Unknowable* and *Exploring Randomness,* in which I explain LISP, and my book *The Limits of Mathematics,* in which I use it. These are technical books.]

MY DIOPHANTINE EQUATION FOR LISP

So you can see how simple LISP is. But actually, it's still not simple enough. In order to make a LISP equation, I had to simplify LISP even more. It's still a powerful language, you can still calculate anything you could before, but the fact that some things are missing and you have to program them yourself starts to become really annoying. It starts to feel more like a machine language and less like a higher-level language. But I can actually put together the entire equation for running LISP, more precisely, the equation for evaluating LISP expressions. Here it is!

Exponential Diophantine Equation Computer:

$$L(k, n, x, y, z, \dots) = R(k, n, x, y, z, \dots)$$

LISP expression k
Value of LISP expression n
Time $x, y, z, \dots$

Two hundred page equation, 20,000 unknowns!

If the LISP expression k has no value, then this equation will have no solution. If the LISP expression k has a value, then this equation will have **exactly one** solution. In this unique solution, n = the value of the expression k.

(Chaitin, *Algorithmic Information Theory,* 1987.)

Guess what, I'm not going to actually show it to you here! Anyway, this is a diophantine equation that's a LISP interpreter. So that's a fairly serious amount of computational ability packed into one diophantine equation.

We're going to put this computer equation to work for us later, in Chapter Six, in order to show that Ω's randomness also infects number theory.

CONCLUDING REMARKS

Some concluding remarks. I've tried to share a lot of beautiful mathematics with you in this chapter. It's math that I'm very passionate about, math that I've spent my life on!

Euclid's proof is classical, Euler's proof is modern, and my proof is post-modern! Euclid's proof is perfect, but doesn't seem to particularly lead anywhere. Euler's proof leads to Riemann and analytic number theory and much extremely technical and difficult modern work. My proof leads to the study of program-size complexity (which actually preceded this proof). The work on Hilbert's 10th shows a marvelously unexpected application of the notion of a computably enumerable set.

I'm not really that interested in the primes. The fact that they now have applications in cryptography makes me even less interested in the primes. What is really interesting about the primes is that there are still lots of things that we don't know about them, it's the philosophical significance of this bizarre situation. What's really interesting is that even in an area of math as simple as this, you immediately get to questions that nobody knows how to answer and things begin to look random and haphazard!

Is there always some structure or pattern there waiting to be discovered? Or is it possible that some things really are lawless, random, patternless—even in pure math, even in number theory?

Later I'll give arguments involving information, complexity and the extremely fundamental notion of irreducibility that strongly suggest that the latter is in fact the case. In fact, that's the *raison d'être* for this book.

Based on his own rather individualistic viewpoint—which happens to be entirely different from mine—Wolfram gives many interesting examples that **also** strongly suggest that there are lots of unsolvable problems in mathematics. I think that his book on *A New Kind of Science* is extremely interesting and provides further evidence that lots of things are unknowable, that the problem is really serious. Right away in mathematics, in 2000 year-old math, you get into trouble, and there seem to be limits to what can be known. Epistemology, which deals with what we can know and why, is the branch of philosophy that worries about these questions. So this is really a book about epistemology, and so is Wolfram's book. We're actually working on philosophy as well as math and physics!

To summarize, in this chapter we've seen that the computer is a powerful new mathematical concept illuminating many questions in mathematics and metamathematics. And we've seen that while formal axiomatic systems are a failure, formalisms for computing are a brilliant success.

To make further progress on the road to Ω, we need to add more to the stew. We need the idea of digital information—that's measured by the size of computer programs—and we also need the idea of irreducible digital information, which is a kind of randomness. In the next chapter we'll discuss the sources of these ideas. We'll see that the idea of complexity comes from biology, the idea of digital information comes from computer software, and the idea of irreducibility—that's my particular contribution—can be traced back to Leibniz in 1686.

Three

DIGITAL INFORMATION: DNA/SOFTWARE/LEIBNIZ

> "Nothing is more important than to see the sources of invention which are, in my opinion, more interesting than the inventions themselves."
>
> —*Leibniz,* as quoted in Pólya, *How to Solve It.*

In this chapter I'm going to show you the "sources of invention" of the ideas of digital information, program-size complexity, and algorithmic irreducibility or randomness. In fact, the genesis of these ideas can be traced to DNA, to software, and to Leibniz himself.

Basically the idea is just to measure the size of software in bits, that's it! But in this chapter I want to explain to you just why this is so important, why this is such a universally applicable idea. We'll see that it plays a key role in biology, where DNA is that software, and in computer technology as well of course, and even in Leibniz's metaphysics where he analyzes what's a law of nature, and how can we decide if something is lawless or not. And later, in Chapter Six, we'll see that it plays a key role in metamathematics, in discussing what a FAS can or cannot achieve.

Let's start with Leibniz.

WHO WAS LEIBNIZ?

Let me tell you about Leibniz.

Leibniz invented the calculus, invented binary arithmetic, a superb mechanical calculator, clearly envisioned symbolic logic, gave the name to topology (analysis situs) and combinatorics, discovered Wilson's theorem (a primality test; see Dantzig, *Number, The Language of Science*), etc. etc. etc.

Newton was a great physicist, but he was definitely inferior to Leibniz both as a mathematician and as a philosopher. And Newton was a rotten human being—so much so that Djerassi and Pinner call their recent book *Newton's Darkness.*

Leibniz invented the calculus, published it, wrote letter after letter to continental mathematicians to explain it to them, initially received all the credit for this from his contemporaries, and then was astonished to learn that Newton, who had never published a word on the subject, claimed that Leibniz had stolen it all from him. Leibniz could hardly take Newton seriously!

But it was Newton who won, not Leibniz.

Newton bragged that he had destroyed Leibniz and rejoiced in Leibniz's death after Leibniz was abandoned by his royal patron, whom Leibniz had helped to become the king of England. It's extremely ironic that Newton's incomprehensible *Principia*—written in the style of Euclid's *Elements*—was only appreciated by continental mathematicians **after** they succeeded in translating it into that effective tool, the infinitesimal calculus that Leibniz had taught them!

Morally, what a contrast! Leibniz was such an elevated soul that he found good in all philosophies: Catholic, Protestant, Cabala, medieval scholastics, the ancients, the Chinese . . . It pains me to say that Newton enjoyed witnessing the executions of counterfeiters he pursued as Master of the Mint.

[The science-fiction writer Neal Stephenson has recently published a trilogy about Newton versus Leibniz, and comes out strongly on Leibniz's side. See also Isabelle

Stengers, *La Guerre des sciences aura-t-elle lieu?,* a play about Newton vs. Leibniz, and the above mentioned book, consisting of two plays and a long essay, called *Newton's Darkness.*]

Leibniz was also a fine physicist. In fact, Leibniz was good at everything. For example, there's his remark that "music is the unconscious joy that the soul experiences on counting without realizing that it is counting." Or his effort to discern prehistoric human migration patterns through similarities between languages—something that is now done with DNA!

So you begin to see the problem: Leibniz is at too high an intellectual level. He's too difficult to understand and appreciate. In fact, you can only really appreciate Leibniz if you are at **his** level. You can only realize that Leibniz has anticipated you **after** you've invented a new field by yourself—which, as C. MacDonald Ross says in his little Oxford University Press book *Leibniz,* has happened to many people.

In fact, that's what happened to me. I invented and developed my theory of algorithmic information, and one day not so long ago, when asked to write a paper for a philosophy congress in Bonn,* I went back to a little 1932 book by Hermann Weyl, *The Open World,* Yale University Press. I had put it aside after being surprised to read in it that Leibniz in 1686 in his *Discours de Métaphysique*—that's the original French, in English, *Discourse on Metaphysics*—had made a key observation about complexity and randomness, **the** key observation that started me on all this at age 15!

[Actually, Weyl himself was a very unusual mathematician, Hilbert's mathematical heir and philosophical opponent, who wrote a beautiful book on philosophy that I had read as a teenager: *Philosophy of Mathematics and Natural Science,* Princeton University Press, 1949. In that book Weyl also recounts Leibniz's idea, but the wording is not as sharp, it's not formulated as clearly and as dramatically as in Weyl's 1932 Yale book. Among his other works: an important book on relativity, *Raum-Zeit-Materie* (space, time, matter).]

"Could Leibniz have done it?!" I asked myself. I put the matter aside until such time as I could check what he had actually said. Years

*This dense Bonn paper is included here as Appendix II. See also the introductory essay, originally published in *American Scientist,* reproduced in Appendix I.

passed . . . Well, for my Bonn paper I finally took the trouble to obtain an English translation of the *Discours,* and then the original French. And I tried to find out more about Leibniz.

It turns out that Newton wasn't the only important opponent that Leibniz had had. You didn't realize that math and philosophy were such dangerous professions, did you?!

The satirical novel *Candide* by Voltaire, which was made into a musical comedy when I was a child, is actually a caricature of Leibniz. Voltaire was Leibniz's implacable opponent, and a terrific Newton booster—his mistress la Marquise du Châtelet translated Newton's *Principia* into French. Voltaire was against one and in favor of the other, not based on an understanding of their work, but simply because Leibniz constantly mentions God, whereas Newton's work seems to fit in perfectly with an atheist, mechanistic worldview. This was leading up to the French revolution, which was against the Church just as much as it was against the Monarchy.

Poor Voltaire—if he had read Newton's private papers, he would have realised that he had backed the wrong man! Newton's beliefs were primitive and literal—Newton computed the age of the world based on the Bible. Whereas Leibniz was never seen to enter a church, and his notion of God was sophisticated and subtle. Leibniz's God was a logical necessity to provide the initial complexity to create the world, and is required because nothing is necessarily simpler than something. That's Leibniz's answer to the question, "Why is there something rather than nothing? For nothing is **simpler** and easier than something" (*Principles of Nature and Grace,* 1714, Section 7).

In modern language, this is like saying that the initial complexity of the universe comes from the choice of laws of physics and the initial conditions to which these laws apply. And if the initial conditions are simple, for example an empty universe or an exploding singularity, then all the initial complexity comes from the laws of physics.

The question of where all the complexity in the world comes from continues to fascinate scientists to this day. For example, it's the focus of Wolfram's book *A New Kind of Science.* Wolfram solves the problem of complexity by claiming that it's only an illusion, that the world is actually very simple. For example, according to Wolfram all the randomness in the world is only pseudo-randomness generated by simple algorithms! That's certainly a philosophical possibility, but does it apply to **this** world? Here it seems that quantum-mechanical randomness provides an inexhaustible source of potential complexity, for example via "frozen accidents," such as biological mutations that change the course of evolutionary history.

Before explaining to what extent and how Leibniz anticipated the starting point for my theory, let me recommend some good sources of information about Leibniz, which are not easy to find. On Leibniz's mathematical work, see the chapter on him in E. T. Bell's *Men of Mathematics* and Tobias Dantzig's *Number, The Language of Science.* On Leibniz the philosopher, see C. MacDonald Ross, *Leibniz,* and *The Cambridge Companion to Leibniz,* edited by Nicholas Jolley. For works **by** Leibniz, including his *Discourse on Metaphysics* and *Principles of Nature and Grace,* see G. W. Leibniz, *Philosophical Essays,* edited and translated by Roger Ariew and Daniel Garber.

Actually, let me start by telling you about Leibniz's discovery of binary arithmetic, which in a sense marks the very beginning of information theory, and then I'll tell you what I discovered in the *Discours de Métaphysique.*

LEIBNIZ ON BINARY ARITHMETIC AND INFORMATION

As I said, Leibniz discovered base-two arithmetic, and he was extremely enthusiastic about it. About the fact that

$$10110 = 1 \times 2^4 + 0 \times 2^3 + 1 \times 2^2 + 1 \times 2^1 + 0 \times 2^0$$
$$= 16 + 4 + 2 = 22\ ???$$

Yes indeed, he sensed in the 0 and 1 bits the combinatorial power to create the entire universe, which is exactly what happens in modern digital electronic computers and the rest of our digital information technology: CDs, DVDs, digital cameras, PCs, . . . It's all 0's and 1's, that's our entire image of the world! You just combine 0's and 1's, and you get everything.

And Leibniz was very proud to note how easy it is to perform calculations with binary numbers, in line with what you might expect if you have reached the logical bedrock of reality. Of course, that observation was also made by computer engineers three centuries later; the idea is the same, even though the language and the cultural context in which it is formulated is rather different.

Here is what Dantzig has to say about this in his *Number, The Language of Science:*

> It is the mystic elegance of the binary system that made Leibniz exclaim: *Omnibus ex nihil ducendis sufficit unum.* (One suffices to derive all out of nothing.) [In German: "Einer hat alles aus nichts gemacht." Word for word: "One has all from nothing made."] Says Laplace:
>
> > Leibniz saw in his binary arithmetic the image of Creation . . . He imagined that Unity represented God, and Zero the void; that the Supreme Being drew all beings from the void, just as unity and zero express all numbers in his system of numeration . . . I mention this merely to show how the prejudices of childhood may cloud the vision even of the greatest men! . . .
>
> Alas! What was once hailed as a monument to monotheism ended in the bowels of a robot. For most of the high-speed calculating machines [computers] operate on the *binary* principle.

In spite of the criticism by Laplace, Leibniz's vision of creating the world from 0's and 1's refuses to go away. In fact, it has begun to inspire some contemporary physicists, who probably have never even heard of Leibniz.

Opening the 1 January 2004 issue of the prestigious journal *Nature,* I discovered a book review entitled "The bits that make up the Universe." This turned out to be a review of von Baeyer, *Information: The New Language of Science* by Michael Nielsen, himself coauthor of the large and authoritative Nielsen and Chuang, *Quantum Computation and Quantum Information.* Here's what Nielsen had to say:

> What is the Universe made of? A growing number of scientists suspect that information plays a fundamental role in answering this question. Some even go as far as to suggest that information-based concepts may eventually fuse with or replace traditional notions such as particles, fields and forces. The Universe may literally be made of information, they say, an idea neatly encapsulated in physicist John Wheeler's slogan: "It from bit" [Matter from information!] . . . These are speculative ideas, still in the early days of development . . . Von Baeyer has provided an accessible and engaging overview of the emerging role of information as a fundamental building block in science.

So perhaps Leibniz was right after all! At any rate, it certainly is a grand vision!

LEIBNIZ ON COMPLEXITY AND RANDOMNESS

Okay, it's time to look at Leibniz's *Discourse on Metaphysics*! Modern science was really just beginning then. And the question that Leibniz raises is, how can we tell the difference between a world that is governed by law—one in which science can apply—and a lawless world? How can we decide if science actually works?! In other words how can we distinguish between a set of observations that obeys a mathematical law and one that doesn't?

And to make the question sharper, Leibniz asks us to think about scattering points at random on a sheet of paper, by closing our eyes and stabbing at it with a pen many times. Even then, observes Leibniz, there will always be a mathematical law that passes through those very points!

Yes, this is certainly the case. For example, if the three points are $(x,y) = (a,A), (b,B), (c,C)$, then the following curve will pass through those very points:

$$y = \frac{A(x-b)(x-c)}{(a-b)(a-c)} + \frac{B(x-a)(x-c)}{(b-a)(b-c)} + \frac{C(x-a)(x-b)}{(c-a)(c-b)}.$$

Just plug in a every time you see an x, and you'll see that the entire right-hand side reduces to A. And this also works if you set $x = b$ or $x = c$. Can you figure out how this works and write the corresponding equation for four points? This is called Lagrangian interpolation, by the way.

So there will always be a mathematical law, no matter what the points are, no matter how they were placed at random!

That seems very discouraging. How can we decide if the universe is capricious or if science actually works?

And here is Leibniz's answer: If the law has to be extremely complicated ("fort composée") then the points are placed at random, they're "irrégulier," not in accordance with a scientific law. But if the law is simple, then it's a genuine law of nature, we're not fooling ourselves!

See for yourself; look at Sections V and VI of the *Discours.*

The way that Leibniz summarizes his view of what precisely it is that makes the scientific enterprise possible is this: "God has chosen that which is the most perfect, that is to say, in which at the same time the hypotheses are as simple as possible, and the phenomena are as rich as possible." The job of the scientist, of course, is then to figure out these simplest possible hypotheses.

(Please note that these ideas of Leibniz are much stronger than Occam's razor, because they tell us **why** Occam's razor works, why it is necessary. Occam's razor merely asserts that the simplest theory is best.)

But how can we **measure** the complexity of a law and compare it with the complexity of the data that it attempts to explain? Because it's only a valid law if it's simpler than the data, hopefully much simpler. Leibniz does not answer that question, which bothered Weyl immensely. But Leibniz had all the pieces, he only had to put them together. For he worshiped 0 and 1, and appreciated the importance of calculating machines.

The way I would put it is like this: I think of a scientific theory as a binary computer program for calculating the observations, which are also written in binary. And you have a law of nature if there is compression, if the experimental data is compressed into a computer program that has a smaller number of bits than are in the data that it explains. The greater the degree of compression, the better the law, the more you understand the data.

But if the experimental data cannot be compressed, if the smallest program for calculating it is just as large as it is (and such a theory can always be found, that can always be done, that's so to speak our "Lagrangian interpolation"), then the data is lawless, unstructured, patternless, not amenable to scientific study, incomprehensible. In a word, random, irreducible!

And **that** was the idea that burned in my brain like a red-hot ember when I was 15 years old! Leibniz would have understood it instantaneously!

But working out the details, and showing that mathematics contains such randomness—that's where my Ω number comes in—that's the hard part, that took me the rest of my life. As they say, genius is 1% inspiration and 99% perspiration. And the devil is in the details. And the detail that cost me the most effort was the idea of self-delimiting information, which I'll explain later in this chapter, and without which, as we'll see in Chapter Six, there would in fact be no Ω number.

The ideas that I've been discussing are summarized in the following diagrams:

Leibniz's Metaphysics:
ideas ⟶ **Mind of God** ⟶ universe
The ideas are simple, but the universe is very complicated! (If science applies!) God minimizes the left-hand side, and maximizes the right-hand side.

Scientific Method:

theory ⟶ **Computer** ⟶ data

The theory is concise, the data isn't.
The data is compressed into the theory!

Understanding is compression!
To comprehend is to compress!

Algorithmic Information Theory:

binary program ⟶ **Computer** ⟶ binary output

What's the smallest program that produces a given output?
If the program is concise and the output isn't, we have a theory.

If the output is random, then no compression is possible,
and the input has to be the same size as the output.

And here's a diagram from Chapter Two, which I think of in exactly the same way:

Mathematics (FAS):

axioms ⟶ **Computer** ⟶ theorems

The theorems are compressed into the axioms!

I think of the axioms as a computer program for generating all theorems. I measure the amount of information in the axioms by the size of this program. Again, you want the smallest program, the most concise set of axioms, the minimum number of assumptions (hypotheses) that give you that set of theorems. One difference with the previous diagrams: This program never halts, it keeps generating theorems forever; it produces an infinite amount of output, not a finite amount of output.

And the same ideas sort of apply to biology:

Biology:

DNA → **Pregnancy** → organism

The organism is determined by its DNA!
The less DNA, the simpler the organism.

Pregnancy is decompression
of a compressed message.

So I think of the DNA as a computer program for calculating the organism. Viruses have a very small program, single-celled organisms have a larger program, and multi-celled organisms need an even larger program.

THE SOFTWARE OF LIFE: LOVE, SEX AND BIOLOGICAL INFORMATION

A passionate young couple can make love several times a night, every night. Nature does not care about recreational sex; the reason that men and women romanticize about each other and are attracted and fall in love is so that they will be fooled into having children, even if they think they are trying to avoid conception! They're actually trying very hard to transmit information from the male to the female. Every time a man ejaculates inside a woman he loves, an enormous number of sperm cells, each with half the DNA software for a complete individual, try to fertilize an egg.

Jacob Schwartz once surprised a computer science class by calculating the bandwidth of human sexual intercourse, the rate of information transmission achieved in human lovemaking. I'm too much of a theoretician to care about the exact answer, which anyway depends on details like how you measure the amount of time that's involved, but his class was impressed that the bandwidth that's achieved is quite respectable!

What is this software like? It isn't written in 0/1 binary like computer software. Instead DNA is written in a 4-letter alphabet, the 4 bases that can be each rung of the twisted double-helix ladder that is a DNA molecule. Adenine, A, thymine, T, guanine, G, and cytosine, C, are those four letters. Individual genes, which code for a single protein, are kilobases of information. And an entire human genome is measured in gigabases, so that's sort of like gigabytes of computer software.

Each cell in the body has the same DNA software, the complete genome, but depending on the kind of tissue or the organ that it's in, it runs different portions of this software, while using many basic subroutines that are common to all cells.

Program (10011 ...) ⟶ **Computer** ⟶ Output

DNA (GCTATAGC ...) ⟶ **Development** ⟶ Organism

And this software is highly conservative, much of it is quite ancient: Many common subroutines are shared among fruit flies, invertebrates, mice and humans, so they have to have originated in an ancient common ancestor. In fact, there is surprisingly little difference between a chimp and a human, or even between a mouse and human.

We are not that unique; Nature likes to reuse good ideas. Instead of starting afresh each time, Nature "solves" new problems by patching—that is, slightly modifying or mutating—the solutions to old problems, as the need arises. Nature is a cobbler, a tinkerer. It's much too much work, it's much too expensive, to start over again each time. Our DNA software accumulates by accretion, it's a beautiful patchwork quilt! And our DNA software also includes all those frozen accidents, those mutations due to DNA copying errors or ionizing radiation, which is a possible pathway for quantum uncertainty to be incorporated in the evolutionary record. In a sense this is an amplification mechanism, one that magnifies quantum uncertainty into an effect that is macroscopically visible.

Another example of this constant re-use of biological ideas is the fact that the human embryo is briefly a fish, or, more generally, the fact that the development of the embryo tends to recapitulate the evolutionary history leading to that particular organism. The same thing happens in human software, where old code is hard to get rid of, since much is built on it, and it quickly becomes full of add-ons and patches.

When a couple falls madly in love, what is happening in information-theoretic terms is that they are saying, what nice subroutines the other person has, let's try combining some of her subroutines with some of his subroutines, let's do that right away! That's what it's really all about! Just like a child cannot have a first name followed by the entire father's and mother's names, because then names would double in size each generation and soon become too long to remember (although the Spanish nobility attempted to achieve this!), a child cannot have all the software, all the DNA, from both parents. So a child's DNA consists of a random selection of subroutines, half from one parent, half from the other. You pass only half of your software to each child, just as you can't include your whole name as part of your child's name.

And in my great-grandmother's generation in the old country, women would have a dozen children, most of whom would die before puberty. So you were trying a dozen mixes of DNA subroutines from both parents. (In the Middle Ages, babies weren't even named til they were a year old, since so many of them would die the first year.) Now instead of trying to keep women pregnant all the time, we depend on massive amounts of expensive medical care to keep alive one or two children, no matter how unhealthy they are. While such medical care is wonderful for the individual, the quality of the human gene pool inevitably deteriorates to match the amount of medical care that is available. The more medical care there is, the sicker people become! The massive amounts of medical care become part of the ecology, and people come to depend on it to survive til child-bearing and for their children to survive in turn . . .

Actually, many times a woman will not even realize that she was briefly pregnant, because the embryo was not at all viable and the pregnancy quickly aborted itself. So parents still actually do try lots of different combinations of their subroutines, even if they only have a few children.

We are just beginning to discover just how powerful biological software is. For example, take human life expectancy. Ageing isn't just a finally fatal accumulation of wear and tear. No, death is programming, the life-expectancy of an individual is determined by an internal alarm clock. Just like the way that "apoptosis," the process by which individual cells are ordered to self-destruct, is an intrinsic part of new cell growth and the plasticity in the organism (it must continually tear itself down in order to continually rebuild itself), a human organism systematically starts to self-destruct according to a pre-programmed schedule, no doubt designed to get him or her out of the way and no longer competing for food with children and younger childbearing individuals. (Although grandparents seem to serve a role to help caring for children.) So this is just software! Well, then it should be easy to modify, you just change a few key parameters in your code! And, in fact, people are beginning to think this can be done. For one invertebrate, the nematode worm *C. elegans,* it has **already** been done. (This is work of Cynthia Kenyon.) For humans, maybe in 50 or a hundred years such modification will be common practice, or permanently incorporated into the human genome.

By the way, bacteria may not have different sexes like we do, but they certainly do pass around useful DNA subroutines, which is how antibiotics quickly create superbugs in hospitals. This is called horizontal gene transfer, because the genes go to contemporaries, not to descendants (that's vertical). And this process results in a kind of "bacterial intelligence," in their ability to quickly adapt to a new environment.

WHAT ARE BITS AND WHAT ARE THEY GOOD FOR?

The Hindus were fascinated by large numbers. There is a parable of a mountain of diamond, the hardest substance in the world, one-mile high and one-mile wide. Every thousand years a beautiful golden bird flies over this diamond mountain and sheds a single tear, which wears away a tiny, tiny bit of the diamond. And the time that it will take for these tears to completely wear away the diamond mountain, that immense time, it is but the blink of an eye to one of the gods!

And when I was a small child, like many future mathematicians I was fascinated by large numbers. I would sit in a stairwell of the apartment building where we lived with a piece of paper and a pencil, and I would write down the number 1 and then double it, and then double it again, and again, until I ran out of paper or of patience:

$$2^0 = 1$$
$$2^1 = 2$$
$$2^2 = 4$$
$$2^3 = 8$$
$$2^4 = 16$$
$$2^5 = 32$$
$$2^6 = 64$$
$$2^7 = 128$$
$$2^8 = 256$$
$$2^9 = 512$$
$$2^{10} = 1024$$
$$2^{11} = 2048$$

$$2^{12} = 4096$$
$$2^{13} = 8192$$
$$2^{14} = 16384$$
$$2^{15} = 32768 \ldots$$

A similar attempt to reach infinity is to take a piece of paper and fold it in half, and then fold that in half, and then do it again and again, until the paper is too thick to fold, which happens rather quickly . . .

What does all this have to do with information? Well, the doubling has a lot to do with bits of information, with messages in binary, and with the bits, or **binary digits,** 0 and 1, out of which we can build the entire universe!

One bit of information, a string 1 bit long, can distinguish two different possibilities. Two bits of information, a string 2 bits long, can distinguish four different possibilities. Each time you add another bit to the message, you double the number of possible messages. There are 256 possible different 8-bit messages, and there are 1024 possible different 10-bit messages. So ten bits of information is roughly equivalent to three digits of information, because ten bits can represent 1024 possibilities, while three digits can represent 1000 possibilities.

Two possibilities:

0, 1.

Four possibilities:

00, 01, 10, 11.

Eight possibilities:

000, 001, 010, 011, 100, 101, 110, 111.

Sixteen possibilities:

0000, 0001, 0010, 0011, 0100, 0101, 0110, 0111,
1000, 1001, 1010, 1011, 1100, 1101, 1110, 1111.

So that's what raw information is, that's what raw binary information looks like inside a computer. Everything is in binary, everything is built up out of 0's and 1's. And computers have clocks to synchronize everything, and in a given clock cycle an electrical signal represents a 1, and no signal represents a 0. These two-state systems can be built very reliably, and then everything in a computer is built up out of them.

What do these binary strings, these strings of bits, these binary messages, what do they represent? Well, the bit strings can represent many things. For instance, numbers:

0:	0
1:	1
2:	10
3:	11
4:	100
5:	101
6:	110
7:	111
8:	1000
9:	1001
10:	1010

Or you can use an 8-bit "byte" to represent a single character, using the ASCII character-code, which is used by most computers. For example:

A:	0100 0001
B:	0100 0010
C:	0100 0011
a:	0110 0001
b:	0110 0010
c:	0110 0011
0:	0011 0000
1:	0011 0001
2:	0011 0010
(:	0010 1000
):	0010 1001
blank:	0010 0000

And you can use a string of 8-bit bytes to represent character strings. For example, in the LISP programming language, how do you write the arithmetic expression for 12 plus 24? This is written in LISP as

```
(+ 12 24)
```

This 9-character LISP expression indicates the result of adding the numbers 12 and 24, and its value is therefore the number 36. This LISP expression has 9 characters because there are seven visible characters plus two blanks, and inside a computer it becomes a $9 \times 8 = 72$-bit bit string. And you can easily build up more complicated objects like this. For example, the LISP expression for adding the product of 3 and 4 to the product of 5 and 6 is this:

```
(+ (* 3 4) (* 5 6))
```

This has 19 characters and is therefore represented as a $19 \times 8 = 152$-bit bit string. But this 152-bit bit string could also be interpreted as an extremely large number, one that is 152 bits or, recalling that 10 bits is about 3 digits' worth of information, about $(152/10 = 15) \times 3 = 45$ digits long.

So bit strings are neutral, they are pure syntax, but it is a convention regarding semantics that gives them meaning. In other words, bit strings can be used to represent many things: numbers, character strings, LISP expressions, and even, as we all know, color pictures, since these are also in our computers. In each case it is the bit string plus a convention for interpreting them that enables us to determine its meaning, be it a number, a character string, a LISP expression, or a color picture. For pictures one must in fact give the height and width of the picture in pixels, or picture elements (points on a display), and then the red/green/blue intensity at each point, each with 8-bit precision, which is a total of 24 bits per pixel.

WHAT IS BIOLOGICAL INFORMATION?

Here is a specific example that is of great interest to us as human beings. Our genetic information (DNA) is written using an alphabet of 4 symbols:

A, C, G, T

These are the symbols for each of the possible bases at each rung of a DNA double-helix. So each of these bases is exactly 2 bits of information, since two bits enables us to specify exactly $2 \times 2 = 4$ possibilities.

In turn, a triple (codon) of 3 bases "codes for" (specifies) a specific amino acid. So each DNA codon is $3 \times 2 = 6$ bits of information. An individual protein is determined by giving the linear sequence of its amino acids in a portion of the DNA that is called a gene. The gene determines the amino-acid "spinal cord" of each protein. Once synthesized in a ribosome, the protein immediately folds into a complicated three-dimensional shape. This folding process isn't well understood and in the current state of the art requires massive amounts of computation to simulate. It is the complicated geometrical form of the protein that determines its biological activity. For example, enzymes are proteins that catalyze (greatly facilitate and speed up) specific chemical reactions by holding the reagents close to each other in exactly the right way for them to react with each other.

That's the story, roughly speaking, but DNA is actually much more sophisticated than that. For example, some proteins turn other genes on and off; in other words, they control gene "expression." We are dealing here with a programming language that can perform complicated calculations and run through sophisticated sequences of gene expression in response to changes in environmental conditions!

And as I said before, the DNA software of some of our cousin apes and other near-relative mammals is surprisingly similar to our own. DNA subroutines are strongly "conserved"; they are reused constantly across many different species. Many of our basic subroutines are present in much more primitive living beings. They haven't changed much; Nature likes to re-use good ideas.

By the way, a human being has 3 giga-bases, that's 3Gb, of DNA:

Human = 3 giga-bases = 6 giga-bits !!!

The conventional units of biological information are kilo-bases (genes), mega-bases (chromosomes) and giga-bases (entire genomes for organisms). That's thousands, millions and billions of bases, respectively.

COMPRESSING DIGITAL PICTURES AND VIDEO

It is said that a picture is worth a thousand words. Indeed, a thousand-by-thousand-point picture has a million pixels, and each pixel requires 3 bytes = 24 bits to specify the admixture of primary colors. So a thousand by a thousand picture is 3 megabytes or 24 megabits of information. We have all experienced the frustration of having to wait for a large picture to take shape in a web browser.

Digital video requires transmitting many pictures per second to create the illusion of smooth, continuous motion, and this requires extremely high internet bandwidth to work well, say, thirty frames per second. This kind of high-bandwidth connection is now usually only available in-house within organizations, but not across the internet between organizations. It would be nice to be able to distribute HDTV, high-resolution digital video, across the web, but this is unfortunately not practical at this time.

But there isn't really that much information there—you don't really have to transmit all those bits. Compression techniques are widely used to speed up the transmission of digital pictures and video. The idea is to take advantage of the fact that large regions of a photo may be more or less the same, and that successive frames of a video may not differ that much from each other. So instead of transmitting the pictures directly, you just send compact descriptions of the pictures, and then at the receiving end you decompress them and re-create the original pictures. It's sort of like the freeze-dried (desiccated) foods that you can take with you up a mountain. Just add water and heat, and you recreate the original foods, which are mostly water and much heavier than their freeze-dried "compressions."

Why does this compression and decompression work so well? It works so well because pictures are far from random. If each pixel had absolutely no connection with any of its neighboring pixels, and if successive frames of a digital video were totally unconnected, then no compression technique would work. Compression techniques are useless if they are applied to noise, to the mad jumble that you get if an antenna is disconnected, because there is absolutely no pattern there to compress away.

Another way to put it is that the most informative picture is one in which each pixel is a complete surprise. Fortunately real pictures are almost never like that. But we shall re-encounter this important theme later on, in Chapter Five, when we'll take incompressibility and use it as the basis for a mathematical definition of a random real number. And then, in Chapter Six, we'll discover such a random number in pure mathematics: the halting probability Ω. Ω is an infinite sequence of bits in which there is no pattern, and there are no correlations. Its bits are mathematical facts that cannot be compressed into axioms that are more concise than they are. So, surprisingly enough, digital TV and DVD compression/decompression technology may actually have something to do with more important questions, namely with philosophy and the limits of knowledge!

Indeed, the philosophical, mathematical ideas antedated these practical applications of compression and decompression. There were no

DVDs when I started working on these ideas in the early 1960's. At that time audio and video recording was analog, not digital. And the cheap computing power necessary to compress and decompress audio and video was simply not available back then. Vinyl records used to use the depth of a groove to record a sound, not 0's and 1's like a CD does. I even remember my grandparents' totally mechanical record player, absolutely without electricity, that still worked perfectly when I was a child on their old, heavy 78-rpm records. What astonishing progress!

Too bad that human society hasn't progressed at the same pace that this technology has! Indeed, in some ways we've retrogressed, we haven't progressed at all. Technology can be improved easily enough, but the human soul, that's immensely hard to improve!

But according to John Maynard Smith and Eörs Szathmáry (see their books *The Major Transitions in Evolution* and *The Origins of Life*), this kind of technological progress isn't important just in consumer electronics, it's also responsible for the major steps forward in the evolution of life on this planet. It played an extremely important role in the creation of vastly new and improved life-forms, which are in a sense multiple **origins** of life, moments when the life on this planet successfully re-invented itself. To be more specific, they see the major steps forward in evolution as occurring at each point in time when biological organisms were able to invent and take advantage of radically new and improved techniques for representing and storing biological information.

And this even extends to human society and to human history! In his books *The New Renaissance* and *Phase Change,* Douglas Robertson makes a convincing case that the major steps forward in human social evolution were because of the invention of language, which separates us from the apes, because of written language, which makes civilization possible, because of the printing press and cheap paper and cheap books, which provoked the Renaissance and the Enlightenment, and because of the PC and the web, which are the current motor for drastic social change. Each major transition occurred because it became possible for human society to record and transmit a vastly increased amount of information. That may not be the **whole** story, but the amount of information that could be recorded and remembered, the availability of information, certainly played a major role.

WHAT IS SELF-DELIMITING INFORMATION?

Okay, so far we've discussed what information is, and the important role that it plays in biology—and even in determining our form of human social organization!—and the idea of compression and decompression. But there's another important idea that I'd like to tell you about right away, which is how to make information be "self-delimiting." What the heck is that?!

The problem is very simple: How can we tell where one binary message ends and another begins, so that we can have **several** messages in a row and not get confused? (This is also, it will turn out, an essential step in the direction of the number Ω!) Well, if we know exactly how long each message is, then there is no problem. But how can we get messages to indicate how long they are?

Well, there are many ways to do this, with increasing degrees of sophistication.

For example, one way to do it is to double each bit of the original message, and then tack on a pair of unequal bits at the end. For example, the plain ordinary binary string

Original string: 011100

becomes the self-delimiting binary string

With bits doubled: 00 11 11 11 00 00 01.

The only problem with this technique is that we've just doubled the length of every string! True, if we see two such strings in a row, if we are given two such strings in a row, we know just where to separate them. For example,

Two self-delimiting strings: 00 11 11 11 00 00 01   11 00 11 00 10

gives us the two separate strings

011100,   1010.

But we had to double the size of everything! That's too big a price to pay! Is there a better way to make information self-delimiting? Yes, there certainly is!

Let's start with

011100

again. But this time, let's put a prefix in front of this string, a header, that tells us precisely how many bits there are in this message (6). To do this, let's write 6 in binary, which gives us 110 (4 + 2), and then let's use the bit-doubling trick on the prefix so that we know where the header ends and the actual information begins:

With a header: 11 11 00 01   011100

In this particular case, using a bit-doubled header like this is no better than doubling the entire original message. But if our original message was very long, doubling the header would be a lot more economical than doubling everything.

Can we do even better?! Yes, we can; there's no need to double the entire header. We can just put a prefix in front of the header that tells us how long the header is, and just double **that.** So now we have **two** headers, and only the first one is doubled, nothing else is. How many bits are there in our original header? Well, the header is 6 = 110, which is only 3 = 11 bits long, which doubled gives us 11 11 01. So now we've got this:

With two headers: 11 11 01   110   011100

In this case using two headers is actually longer than just using one header. But if our original message were very, very long, then this would save us a lot of bits.

And you can keep going on and on in this way, by adding more and more headers, each of which tells us the size of the **next** header, and only the very **first** header is doubled!

So I think you get the idea. There are lots and lots of ways to make information self-delimiting! And the number of bits that you have to add in order to do this isn't very large. If you start with an N-bit message, you can make it self-delimiting by adding only the number of bits you need to specify N in a self-delimiting manner. That way you know where all the header information, the part that tells us N, ends, and where the actual information, all N bits of it, begins. Okay? Is that clear?

This is more important than it seems. First of all, self-delimiting information is **additive,** which means that the number of bits it takes to convey two messages in a row is just the sum of the number of bits it takes to convey each message separately. In other words, if you think of individual messages as subroutines, you can combine many subroutines into a larger program without having to add any bits—as long as you know just how many subroutines there are, which you usually do.

And that's not all. We'll come back to self-delimiting information later when we discuss how the halting probability Ω is defined, and why the definition works and actually defines a probability. In a word, it works because we stipulate that these programs, the ones that the halting probability Ω counts, have to be self-delimiting binary information. Otherwise Ω turns out to be nonsense, because you can't use a single number to count the programs of **any** size that halt, you can only do it for programs of a particular size, which isn't very interesting.

We'll discuss this again later. It's probably the most difficult thing to understand about Ω. But even if I can't explain it to you well enough for you to understand it, just remember that we're always dealing with self-delimiting information, and that all the computer programs that we've been and will be considering are self-delimiting binary; that'll give you the general idea of how things work. Okay?

Algorithmic Information Theory:

self-delimiting information → **Computer** → output

What's the smallest self-delimiting program that produces a given output? The size of that program in bits is the complexity *H*(output) of the output.

After all, you don't want to become an expert in this field (at least not yet!), you just want to get a general idea of what's going on, and of what the important ideas are.

MORE ON INFORMATION THEORY & BIOLOGY

On the one hand, different kinds of organisms need different amounts of DNA. And the DNA specifies how to construct the organism, and how it works.

DNA → **Development/Embryogenesis/Pregnancy** → Organism

Complexity/Size Measure:
kilo/mega/gigabases of DNA.

And one can more or less classify organisms into a complexity hierarchy based on the amount of DNA they need: viruses, cells without nuclei, cells with nuclei, multi-celled organisms, humans . . .

That's the biological reality. Now let's abstract a highly-simplified mathematical model of this:

Program ⟶ **Computer** ⟶ Output

Complexity/Size Measure:
bits/bytes kilo/mega/gigabytes of software.

H(Output) = size of smallest program for computing it.

Actually, a byte = 8 bits, and kilo/mega/gigabytes are the conventional units of software size, leading to confusion with bases (= 2 bits), which also starts with the letter "b"! One way to avoid the confusion is to use capital "B" for "bytes," and lowercase "b" for "bases." So KB, MB, GB, TB = kilo/mega/giga/terabytes, and Kb, Mb, Gb, Tb = kilo/mega/giga/terabases. And kilo = one thousand (10^3), mega = one million (10^6), giga = one billion (10^9), tera = a million millions (10^{12}). However, as a theoretician, I'll think in just plain bits; usually, I won't use any of these practical information size measures.

Now let me be provocative: Science is searching for the DNA for the Universe!

A Mixed-Up Model:

DNA ⟶ **Computer** ⟶ Universe!

What is the complexity of the entire universe?
What's the smallest amount of DNA/software
to construct the world?

So, in a way, AIT is inspired by biology! Certainly biology is the domain of the complex, it's the most obvious example of complexity in the world, not physics, where there are simple unifying principles, where there are simple equations that explain everything! But can AIT contribute to biology?! Can ideas flow from AIT to biology, rather than from biology to AIT? One has to be very careful!

First of all, the

Algorithmic Information Theory:

program ⟶ **Computer** ⟶ output

What's the smallest program for a given output?

model of AIT imposes **no** time limit on the computation. But in the world of biology, 9 months is already a long time to produce a new organism, that's a long pregnancy. And some genes are repeated because one needs large quantities of the protein that they code for. So AIT is much too simple-minded a model to apply to biology. DNA is **not** the minimum size for the organism that it produces.

In my theory a program should have no redundancy. But in DNA and the real world, redundancy is good. If there were none, any change in a message yields **another** valid message. But redundancy makes it possible to do error correction and error detection, which is very important for DNA (indeed DNA actually contains **complementary base pairs,** not individual bases, so that is already a form of redundancy).

But unfortunately, in order for us to be able to prove theorems, we need to use a less complicated model, a toy model, one that applies well to metamathematics, but that does not apply well to biology. Remember: Pure mathematics is much easier to understand, much simpler, than the messy real world!

Complexity!

It's an idea from Biology imported into Mathematics.

*It's **not** an idea from Mathematics imported into Biology.*

AIT also fails in biology in another crucial respect. Look at a crystal and at a gas. One has high program-size complexity, the other has low program-size complexity, but **neither** is organized, neither is of any biological interest!

Gas, Crystal

H(Crystal) is very low because
it's a regular array of motionless atoms.

H(Gas) is very high because you have to specify
where each atom is and where it is going and how fast.

Here, however, is a (very theoretical) biological application of AIT, using mutual information, the extent to which two things considered together are simpler than considered apart. In other words, this is the extent to which the smallest program that computes **both** simultaneously is smaller than the sum of the size of the individual smallest programs for each separately.

Mutual Information Between X and Y

The mutual information is equal to $H(X) + H(Y) - H(X, Y)$.

It's **small** if $H(X, Y)$ is approximately equal to $H(X) + H(Y)$.

It's **large** if $H(X, Y)$ is much smaller than $H(X) + H(Y)$.

Note that $H(X, Y)$ **cannot** be larger than $H(X) + H(Y)$ because programs are self-delimiting binary information.

If two things have very little in common, it makes no difference if we consider them together or separately. But if they have a lot in common, then we can eliminate common subroutines when we calculate them at the same time, and so the mutual information will be large. How about biology? Well, there is the problem of the whole versus the parts. By what right can we partition the world of our experience into parts, instead of considering it as a unified whole? I'll discuss this in a moment. Mutual information also has a very theoretical application in music theory, not just in theoretical biology. How?

In music, we can use this measure of mutual information to see how close two compositions are, to see how much they have in common. Presumably, two works by Bach will have higher mutual information than a work by Bach and one by Shostakovich. Or compare the entire body of work of two composers. Two Baroque composers should have more in common than a Baroque and a Romantic composer.

Bach, Shostakovich

H(Bach, Shostakovich) is approximately equal to H(Bach) + H(Shostakovich).

Bach_1 = Brandenburg Concertos,
Bach_2 = Art of the Fugue

$H(\text{Bach}_1, \text{Bach}_2)$ is much less than $H(\text{Bach}_1) + H(\text{Bach}_2)$.

How about biology? As I said, there is the problem of the whole versus the parts. By what right can we partition the universe into parts, rather than consider it as an interacting whole? Well, the parts of an organism have high mutual information. (Because they all contain the full genome for that organism, they all have the same DNA, even though different cells in different tissues turn on different parts of that DNA.) Also, it is natural to divide a whole into pieces if the program-size complexity decomposes additively, that is, if the complexity of the whole is approximately equal to the sum of the complexity of those parts, which means that their mutual interactions are not as important as their internal interactions. This certainly applies to living beings!

Fred, Alice

H(Fred, Alice) is approximately equal to H(Fred) + H(Alice).

Fred's left arm, Fred's right arm

H(left arm, right arm) is much less than H(left arm) + H(right arm).

But do we really have the right to talk about the complexity of a physical object like H(Alice), H(Fred)?! Complexity of digital objects, yes, they're all just finite strings of 0's and 1's. But physicists usually use **real numbers** to describe the physical world, and they assume that space and time and many measurable quantities can change continuously. And a real number isn't a digital object, it's an analog object. And since it varies continuously, not in discrete jumps, if you convert such a number into bits you get an infinite number of bits. But computers can't perform computations with numbers that have an infinite number of bits! And my theory of algorithmic information is based on what computers can do.

Does this pull the rug out from under everything? Fortunately it doesn't.

In Chapter Four we'll discuss whether the physical world is really continuous or whether it might actually be discrete, like some rebels are beginning to suspect. We'll see that there are in fact plenty of physical reasons to reject real numbers. And in Chapter Five we'll discuss mathematical and philosophical reasons to reject real numbers. There are lots of arguments against real numbers, it's just that people usually aren't willing to listen to any of them!

So our digital approach may in fact have a truly broad range of applicability. After this reassuring vindication of our digital viewpoint, the road will finally be clear for us to get to Ω, which will happen in Chapter Six.

Four

INTERMEZZO

THE PARABLE OF THE ROSE

Let's take a break. Please take a look at this magnificent rose:

In his wonderfully philosophical short story *The Rose of Paracelsus* (1983), Borges depicts the medieval alchemist Paracelsus, who had the fame of being able to re-create a rose from its ashes:

> The young man raised the rose into the air.
>
> "You are famed," he said, "for being able to burn a rose to ashes and make it emerge again, by the magic of your art. Let me witness that prodigy. I ask that of you, and in return I will offer up my entire life." . . .
>
> "There is still some fire there," said Paracelsus, pointing toward the

hearth. "If you cast this rose into the embers, you would believe that it has been consumed, and that its ashes are real. I tell you that the rose is eternal, and that only its appearances may change. At a word from me, you would see it again."

"A word?" the disciple asked, puzzled. "The furnace is cold, and the retorts are covered with dust. What is it you would do to bring it back again?"[6]

Now let's translate this into the language of our theory.

The *algorithmic information content* (program-size complexity) H(rose) of a rose is defined to be the size in bits of the smallest computer program (algorithm) p_{rose} that produces the digital image of this rose.

This minimum-size algorithmic description p_{rose} captures the **irreducible essence** of the rose, and is the number of bits that you need to preserve in order to be able to recover the rose:

ashes ⟶ **Alchemy/Paracelsus** ⟶ rose

minimum-size program p_{rose} ⟶ **Computer** ⟶ rose

size in bits of p_{rose} =

algorithmic information content H(rose) of the rose

H(rose) measures the **conceptual complexity** of the rose, that is, the difficulty (in bits) of conceiving of the rose, the number of **indepen-**

[6]**Source:** Jorge Luis Borges, *Collected Fictions,* translated by Andrew Hurley, Penguin Books, 1999, pp. 504–507. For the original *La rosa de Paracelso,* see Borges, *Obras Completas,* Tomo III, Emecé, 1996, pp. 387–390.

dent yes/no choices that must be made in order to do this. The larger *H*(rose) is, the **less inevitable** it is for God to create this particular rose, as an independent act of creation.

In fact, the work of Edward Fredkin, Tommaso Toffoli and Norman Margolus on reversible cellular automata has shown that there are discrete toy universes in which **no information is ever lost,** that is, speaking in medieval terms, **the soul is immortal.** Of course, this is not a terribly personal kind of immortality; it only means that enough information is always present to be able to reverse time and recover any previous state of the universe.

Furthermore the rose that we have been considering is only a jpeg image, it's only a digital rose. What about a real rose? Is that analog or digital? Is physical reality analog or digital? In other words, in this universe are there infinite precision real numbers, or is everything built up out of a finite number of 0's and 1's?

Well, let's hear what Richard Feynman has to say about this.

> It always bothers me that, according to the laws as we understand them today, it takes a computing machine an infinite number of logical operations to figure out what goes on in no matter how tiny a region of space, and no matter how tiny a region of time. How can all that be going on in that tiny space? Why should it take an infinite amount of logic to figure out what one tiny piece of space/time is going to do? So I have often made the hypothesis that ultimately physics will not require a mathematical statement, that in the end the machinery will be revealed, and the laws will turn out to be simple like the chequer board . . .
>
> —Feynman, *The Character of Physical Law.*

Fredkin told me that he spent years trying to make Feynman take digital physics seriously, and was very pleased to see his ideas reproduced in this passage in Feynman's book! For an idea to be successful, you have to give it away, you have to be willing to let other people think that it was theirs! You can't be possessive, you can't be jealous! . . .

I used to be fascinated by physics, and I wanted very badly to be able to know if physical reality is discrete or continuous.

But what I really want to know now is: "What is life?," "What is mind?," "What is intelligence?," "What is consciousness?," "What is creativity?" By working in a toy world and using digital physics, we can hopefully avoid getting bogged down in the physics and we can concentrate on the biology instead.

I call this theoretical theoretical physics, because even if these models do not apply to **this particular** world, they are **interesting possible** worlds!

And my ultimate ambition, which hopefully somebody will someday achieve, would be to **prove** that life, intelligence and consciousness must with high probability evolve in one of these toy worlds. In playing God like this, it's important to get out more than you put in, because otherwise the whole thing can just be a cheat. So what would really be nice is to be able to obtain **more** intelligence, or a higher degree of consciousness, than you yourself possess! You, who set up the rules of the toy universe in the first place.

This is certainly not an easy task to achieve, but I think that this kind of higher-level understanding is probably more fundamental than just disentangling the microstructure of the physical laws of this one world.

THEORETICAL PHYSICS & DIGITAL PHILOSOPHY

Nevertheless, there are some intriguing hints that this particular universe may in fact be a discrete digital universe, not a continuous analog universe the way that most people would expect.

In fact these ideas actually go back to Democritus, who argued that matter must be discrete, and to Zeno, who even had the audacity to suggest that continuous space and time were self-contradictory impossibilities.

Through the years I've noticed many times, as an armchair physicist, places where physical calculations diverge to infinity at extremely small distances. Physicists are adept at not asking the wrong question, one that gives an infinite answer. But I'm a mathematician, and each time I would wonder if Nature wasn't really trying to tell something, namely that real numbers and continuity are a sham, and that infinitesimally small distances **do not exist!**

Two Examples: the infinite amount of energy stored in the field around a point electron according to Maxwell's theory of electromagnetism, and the infinite energy content of the vacuum according to quantum field theory.

In fact, according to quantum mechanics, infinitely precise measurements require infinite energies (infinitely big and expensive atom smashers), but long before that you get gravitational collapse into a black hole, if you believe in general relativity theory.

For more on ideas like this, see the discussion of the Bekenstein bound and the holographic principle in Lee Smolin, *Three Roads to Quantum Gravity.*

And although it's not usually presented that way, string theory was invented to eliminate these divergences as distances become arbitrarily small. String theory does this by eliminating point field sources like the electron, which as I pointed out causes difficulties for Maxwell's theory of electromagnetism. In string theory elementary particles are changed from points into loops of string. And string theory provides a crucial **minimum distance scale,** which is the length of these loops of string. (They're in a tightly curled-up extra dimension.) So you can no longer make distances go below that minimum length scale.

That's one of the reasons that string theory had to be invented, in order to get rid of arbitrarily small distances!

Last but not least, I should share with you the following experience: For many years, whenever I would bump into my esteemed late colleague Rolf Landauer in the halls of the IBM Watson Research Center in Yorktown Heights, he would often gently assure me that no physical quantity has ever been measured with more than, say, twenty digits of precision. And the experiments that achieve that degree of precision are masterpieces of laboratory art and the sympathetic search for physical quantities stable enough and sharply defined enough **to allow themselves** to be measured with that degree of precision, not to mention eliminating all possible perturbing sources of noise! So why, Rolf would ask, should **anyone** believe in arbitrarily precise measurements or in the real numbers used to record the results of such measurements?!

So, you see, there are lots of reasons to suspect that we might live in a digital universe, that God prefers to be able to copy things exactly when he has to, rather than to get the inevitable increase in noise that accompanies analog copying!

And in the next chapter I'd like to continue in Zeno's footsteps as well as Rolf's and argue that **a number with infinite precision, a so-called real number, is actually rather unreal!**

Five

THE LABYRINTH OF THE CONTINUUM

The "labyrinth of the continuum" is how Leibniz referred to the philosophical problems associated with real numbers, which we shall discuss in this chapter. So the emphasis here will be on philosophy and mathematics, rather than on physics as in the last chapter.

What is a real number? Well, in geometry it's the length of a line, measured exactly, with infinite precision, for example 1.2749591 … , which doesn't sound too problematical, at least at first. And in analytic geometry you need two real numbers to locate a point (in two dimensions), its distance from the x axis, and its distance from the y axis. One real will locate a point on a line, and the line that we will normally consider will be the so-called "unit interval" consisting of all the real numbers from zero to one. Mathematicians write this interval as [0, 1), to indicate that 0 is included but 1 is not, so that all the real numbers corresponding to these points have no integer part, only a decimal fraction. Actually, [0, 1] works too, as long as you write one as 0.99999 … instead of as 1.00000 … But not to worry, we're going to ignore all these subtle details. You get the general idea, and that's enough for reading this chapter.

[By the way, why is it called "real"? To distinguish it from so-called "imaginary" numbers like $\sqrt{-1}$. Imaginary numbers are neither more nor less imaginary than real numbers, but there was initially, several centuries ago, including at the time of Leibniz, much resistance to placing them on an equal footing with real numbers. In a letter to Huygens, Leibniz points out that calculations that temporarily traverse this imaginary world can in fact start and end with real numbers. The usefulness of such a procedure was, he argued, an argument in favor of such numbers. By the time of Euler, imaginaries were extremely useful. For example, Euler's famous result that

$$e^{ix} = \cos x + i \sin x$$

totally tamed trigonometry. And the statement (Gauss) that an algebraic equation of degree n has exactly n roots only works with the aid of imaginaries. Furthermore, the theory of functions of a complex variable (Cauchy) shows that the calculus and in particular so-called power series

$$a_0 + a_1x + a_2x^2 + a_3x^3 + \ldots$$

make much more sense with imaginaries than without. The final argument in their favor, if any was needed, was provided by Schrödinger's equation in quantum mechanics, in which imaginaries are absolutely essential, since quantum probabilities (so-called "probability amplitudes") have to have direction as well as magnitude.]

As is discussed in Burbage and Chouchan, *Leibniz et l'infini,* PUF, 1993, Leibniz referred to what we call the infinitesimal calculus as "the calculus of transcendentals." And he called curves "transcendental" if they cannot be obtained via an algebraic equation, the way that the circles, ellipses, parabolas and hyperbolas of analytic geometry most certainly can.

Leibniz was extremely proud of his quadrature of the circle, a problem that had eluded the ancient Greeks, but that he could solve with *transcendental* methods:

$$\frac{\pi}{4} = 1 - \frac{1}{3} + \frac{1}{5} - \frac{1}{7} + \frac{1}{9} - \frac{1}{11} + \frac{1}{13} + \ldots$$

What is the quadrature of the circle? The problem is to geometrically construct a square having the same area as a given circle, that is, to determine the area of the circle. Well, that's πr^2, r being the radius of the circle, which converts the problem into determining π, precisely what Leibniz accomplished so elegantly with the infinite series displayed above.

Leibniz could not have failed to be aware that in using this term he was evoking the notion of God's transcendence of all things human, of human limitations, of human finiteness.

As often happens, history has thrown away the philosophical ideas that inspired the creators and kept only a dry technical husk of what they thought that they had achieved. What remains of Leibniz's idea of transcendental methods is merely the distinction between algebraic numbers and transcendental numbers. A real number x is algebraic if it is the solution of an equation of the form

$$ax^n + bx^{n-1} + \ldots + px + q = 0$$

where the constants *a, b,* … are all integers. Otherwise *x* is said to be transcendental. The history of proofs of the existence of transcendental numbers is rich in intellectual drama, and is one of the themes of this chapter.

Similarly, it was Cantor's obsession with God's infiniteness and transcendence that led him to create his spectacularly successful but extremely controversial theory of infinite sets and infinite numbers. What began, at least in Cantor's mind, as a kind of madness, as a kind of mathematical theology full—necessarily full—of paradoxes, such as the one discovered by Bertrand Russell, since any attempt by a finite mind to apprehend God is inherently paradoxical, has now been condensed and desiccated into an extremely technical and untheological field of math, modern axiomatic set theory.

Nevertheless, the intellectual history of the proofs of the existence of transcendental numbers is quite fascinating. New ideas totally transformed our way of viewing the problem, not once, but in fact **five** times! Here is an outline of these developments:

- Liouville, Hermite and Lindemann, with great effort, were the first to exhibit individual real numbers that could be proved to be transcendental. Summary: **individual transcendentals.**
- Then Cantor's theory of infinite sets revealed that the transcendental reals had the same cardinality as the set of all reals, while the algebraic reals were merely as numerous as the integers, a smaller infinity. Summary: **most reals are transcendental.**
- Next Turing pointed out that all algebraic reals are computable, but again, the uncomputable reals are as numerous as the set of all reals, while the computable reals are only as numerous as the integers. The existence of transcendentals is an immediate corollary. Summary: **most reals are uncomputable and *therefore* transcendental.**
- The next great leap forward involves probabilistic ideas: the set of random reals was defined, and it turns out that with probability one, a real number is random and therefore necessarily un-

computable and transcendental. Non-random, computable and algebraic reals all have probability zero. So now you can get a transcendental real merely by picking a real number at random with an infinitely sharp pin, or, alternatively, by using independent tosses of a fair coin to get its binary expansion. Summary: **reals are transcendental/uncomputable/random with probability one.** And in the next chapter we'll exhibit a natural construction that picks out an individual random real, namely the halting probability Ω, without the need for an infinitely sharp pin.

- Finally, and perhaps even more devastatingly, it turns out that the set of all reals that can be individually named or specified or even defined or referred to—constructively or not—within a formal language or within an individual FAS, has probability zero. Summary: **reals are un-nameable with probability one.**

So the set of real numbers, while natural—indeed, immediately given—geometrically, nevertheless remains quite elusive:

Why should I believe in a real number if I can't calculate it, if I can't prove what its bits are, and if I can't even refer to it? And each of these things happens with probability one! The real line from 0 to 1 looks more and more like a Swiss cheese, more and more like a stunningly black high-mountain sky studded with pin-pricks of light.

Let's now set to work to explore these ideas in more detail.

THE "UNUTTERABLE" AND THE PYTHAGOREAN SCHOOL

This intellectual journey actually begins, as is often the case, with the ancient Greeks. Pythagoras is credited with naming both mathematics and philosophy. And the Pythagoreans believed that number—whole numbers—rule the universe, and that God is a mathematician, a point of view largely vindicated by modern science, especially quantum mechanics, in which the hydrogen atom is modeled as a musical instrument that produces a discrete scale of notes. Although, as we saw in Chapter Three, perhaps God is actually a computer programmer!

Be that as it may, these early efforts to understand the universe suffered a serious setback when the Pythagoreans discovered geometrical lengths that cannot be expressed as the ratio of two whole numbers. Such lengths are called irrational or incommensurable. In other words, they discovered real numbers that cannot be expressed as a ratio of two whole numbers.

How did this happen?

The Pythagoreans considered the unit square, a square one unit in length on each side, and they discovered that the size of both of the two diagonals, $\sqrt{2}$, isn't a rational number n/m. That is to say, it cannot be expressed as the ratio of two integers. In other words, there are no integers n and m such that

$$\left(\frac{n}{m}\right)^2 = 2 \qquad \text{or} \qquad n^2 = 2m^2$$

An elementary proof of this from first principles is given in Hardy's well-known *A Mathematician's Apology.* He presents it there because he believes that it's a mathematical argument whose beauty anyone should be able to appreciate. However, the proof that Hardy gives, which is actually from Euclid's *Elements,* does not give as much insight as a more advanced proof using unique factorization into primes. I **did not** prove unique factorization in Chapter Two. Nevertheless, I'll use it here. It is relevant because the two sides of the equation

$$n^2 = 2m^2$$

would give us **two different** factorizations of the same number. How?

Well, factor n into primes, and factor m into primes. By doubling the exponent of each prime in the factorizations of n and m,

$$2^{\alpha}\, 3^{\beta}\, 5^{\gamma} \ldots \longrightarrow 2^{2\alpha}\, 3^{2\beta}\, 5^{2\gamma} \ldots,$$

we get factorizations of n^2 and m^2. This gives us a factorization of n^2 in which the exponent of 2 is even, and a factorization of $2m^2$ in which the exponent of 2 is odd. So we have two different factorizations of the same number into primes, which is impossible.

According to Dantzig, *Number, The Language of Science,* the discovery of irrational or incommensurable numbers like $\sqrt{2}$

> caused great consternation in the ranks of the Pythagoreans. The very name given to these entities testifies to that. *Algon,* the *unutterable,* these incommensurables were called . . . How can number dominate the universe when it fails to account even for the most immediate aspect of the universe, namely *geometry?* So ended the first attempt to exhaust nature by number.

This intellectual history also left its traces in the English language: In English such irrationals are referred to as "surds," which comes from the French *sourd-muet,* meaning deaf-mute, one who cannot hear or speak. So the English word "surd" comes from the French word for "deaf-mute," and algon = mute. In Spanish it's *sordomudo,* deaf-mute.

In this chapter we'll retrace this history, and we'll see that real numbers not only confound the philosophy of Pythagoras, they confound as well Hilbert's belief in the notion of a FAS, and they provide us with many additional reasons for doubting their existence, and for remaining quite skeptical. To put it bluntly, our purpose here is to review and discuss the mathematical arguments **against real numbers.**

THE 1800S: INDIVIDUAL TRANSCENDENTALS (LIOUVILLE, HERMITE, LINDEMANN)

Although Leibniz was extremely proud of the fact that he had been able to square the circle using transcendental methods, the 1800s wanted to be **sure** that they were really required. In other words, they demanded proofs that π and other individual numbers defined via the sums of infinite series **were not** the solution of any algebraic equation.

Finding a natural specific example of a transcendental real turned out to be much harder than expected. It took great ingenuity and cleverness to exhibit provably transcendental numbers!

The first such number was found by Liouville:

$$\text{Liouville number} = \frac{1}{10^{1!}} + \frac{1}{10^{2!}} + \ldots + \frac{1}{10^{n!}} + \ldots$$

He showed that algebraic numbers cannot be approximated that well by rational numbers. In other words, he showed that his number cannot be algebraic, because there are rational approximations that work too well for it: they can get too close, too fast. But Liouville's number isn't a natural example, because no one had ever been interested in this particular number before Liouville. It was constructed precisely so that Liouville could prove its transcendence. What about π and Euler's number *e*?

Euler's number

$$e = 1 + \frac{1}{1!} + \frac{1}{2!} + \frac{1}{3!} + \ldots + \frac{1}{n!} + \ldots$$

was finally proved transcendental by Hermite. Here at last was a natural example! This was an important number that people really cared about!

But what about the number that Leibniz was so proud of conquering? He had squared the circle by transcendental methods:

$$\frac{\pi}{4} = \frac{1}{1} - \frac{1}{3} + \frac{1}{5} - \frac{1}{7} + \frac{1}{9} - \frac{1}{11} + \ldots$$

But can you prove that transcendental methods are really necessary? This question attracted a great deal of attention after Hermite's result, since π seemed to be the obvious next candidate for a transcendence proof. This feat was finally accomplished by Lindemann, provoking the famous remark by Kronecker that "Of what use is your beautiful proof, since π **does not exist!**" Kronecker was a follower of Pythagoras; Kronecker's best known statement is, "God created the integers; all the rest is the work of man!"

These were the first steps on the long road to understanding transcendence, but they were difficult complicated proofs that were specially tailored for each of these specific numbers, and gave no general insight into what was going on.

CANTOR: THE NUMBER OF TRANSCENDENTALS IS A HIGHER ORDER INFINITY THAN THE NUMBER OF ALGEBRAIC REALS

As I've said, a real number is one that can be determined with arbitrary precision, such as $\pi = 3.1415926 \ldots$ Nevertheless, in the late 1800s two mathematicians, Cantor and Dedekind, were moved to come up with much more careful definitions of a real number. Dedekind did it via "cuts," thinking of an irrational real r as a way to partition all the rational numbers n/m into those less than r and those greater than r. In Cantor's case a real was defined as an infinite sequence of rational numbers n/m that approach r more and more closely.[7]

History did not take any more kindly to their work than it has to any other attempt at a "final solution."

But first, let me tell you about Cantor's theory of infinite sets and his invention of new, infinite numbers for the purpose of measuring the sizes of all infinite sets. A very bold theory, indeed!

Cantor's starting point is his notion of comparing two sets, finite or infinite, by asking whether or not there is a one-to-one correspondence, a pairing between the elements of the two sets that exhausts both sets and leaves no element of either set unpaired, and no element of one of the sets paired with more than one partner in the other set. If this can be done, then Cantor declares that the two sets are equally big.

Actually, Galileo had mentioned this idea in one of his dialogues, the one published at the end of his life when he was under house arrest. Galileo points out that there are precisely as many positive integers 1, 2, 3, 4, 5, ... as there are square numbers 1, 4, 9, 16, 25, ... Up to that point, history has decided that Galileo was right on target.

However, he then declares that the fact that the squares are just a

[7]Earlier versions of the work of Dedekind and of Cantor on the reals are due to Eudoxus and to Cauchy, respectively. History repeats itself, even in mathematics.

tiny fraction of all the positive integers contradicts his previous observation that they are equally numerous, and that this paradox precludes making any sense of the notion of the size of an infinite set.

The paradox of the whole being equivalent to one of its parts may have deterred Galileo, but Cantor and Dedekind took it entirely in stride. It did not deter them at all. In fact, Dedekind even put it to work for him, he used it. Dedekind **defined** an infinite set to be one having the property that a proper subset of it is just as numerous as it is! In other words, according to Dedekind, a set is infinite if and only if it can be put in a one-to-one correspondence with a part of itself, one that excludes some of the elements of the original set!

Meanwhile, Dedekind's friend Cantor was starting to apply this new way of comparing the size of two infinite sets to common everyday mathematical objects: integers, rational numbers, algebraic numbers, reals, points on a line, points in the plane, etc.

Most of the well-known mathematical objects broke into two classes: 1) sets like the algebraic real numbers and the rational numbers, which were exactly as numerous as the positive integers, and are therefore called "countable" or "denumerable" infinities, and 2) sets like the points in a finite or infinite line or in the plane or in space, which turned out all to be exactly as numerous as each other, and which are said to "have the power of the continuum." And this gave rise to two new infinite numbers, $\aleph_0$ (aleph-nought) and ***c***, both invented by Cantor, that are, respectively, the size (or as Cantor called it, the "power" or the "cardinality") of the positive integers and of the continuum of real numbers.

Comparing Infinities!

$$\#\{\text{reals}\} = \#\{\text{points in line}\} = \#\{\text{points in plane}\} = \boldsymbol{c}$$

$$\#\{\text{positive integers}\} = \#\{\text{rational numbers}\}$$
$$= \#\{\text{algebraic real numbers}\} = \aleph_0$$

Regarding his proof that there were precisely as many points in a plane as there are in a solid or in a line, Cantor remarked in a letter to Dedekind, "Je le vois, mais je ne le crois pas!," which means "I see it, but I don't believe it!," and which happens to have a pleasant melody in French.

And then Cantor was able to prove the extremely important and basic theorem that $\boldsymbol{c}$ is larger than $\aleph_0$, that is to say, that the continuum is a nondenumerable infinity, an uncountable infinity, in other words, that there are more real numbers than there are positive integers, infinitely more. This he did by using Cantor's well-known diagonal method, explained in Wallace, *Everything and More,* which is all about Cantor and his theory.

In fact, it turns out that the infinity of transcendental reals is exactly as large as the infinity of all reals, and the smaller infinity of algebraic reals is exactly as large as the infinity of positive integers. Immediate corollary: most reals are transcendental, not algebraic, infinitely more so.

Well, this is like stealing candy from a baby! It's much less work than struggling with individual real numbers and trying to prove that they are transcendental! Cantor gives us a much more general perspective from which to view this particular problem. And it's much easier to see that **most** reals are transcendental than to decide if a **particular** real number happens to be transcendental!

So that's the first of what I would call the "philosophical" proofs that transcendentals exist. Philosophical as opposed to highly technical, like flying by helicopter to the top of the Eiger instead of reaching the summit by climbing up its infamous snow-covered north face.

Is it really that easy? Yes, but this set-theoretic approach created as many problems as it solved. The most famous is called Cantor's continuum problem.

What is Cantor's continuum problem?

Well, it's the question of whether or not there happens to be any set that has more elements than there are positive integers, and that has fewer elements than there are real numbers. In other words, is there an infinite set whose cardinality or power is bigger than $\aleph_0$ and smaller than $\boldsymbol{c}$? In other words, is $\boldsymbol{c}$ the next infinite number after $\aleph_0$, which has the name $\aleph_1$ (aleph-one) reserved for it in Cantor's theory, or are there a lot of other aleph numbers in between?

Cantor's Continuum Problem

Is there a set S such that $\aleph_0 < \#S < \boldsymbol{c}$?

In other words, is $\boldsymbol{c} = \aleph_1$, which is the first cardinal number after $\aleph_0$?

A century of work has not sufficed to solve this problem!

An important milestone was the proof by the combined efforts of Gödel and Paul Cohen that the usual axioms of axiomatic set theory (as opposed to the "naive" paradoxical original Cantorian set theory) do not suffice to decide one way or another. You can add a new axiom asserting there is a set with intermediate power, or that there is no such set, and the resulting system of axioms will not lead to a contradiction (unless there was already one there, without even having to use this new axiom, which everyone fervently hopes is not the case).

Since then there has been a great deal of work to see if there might be new axioms that set theorists can agree on that might enable them to settle Cantor's continuum problem. And indeed, something called the axiom of projective determinacy has become quite popular among set theorists, since it permits them to solve many open problems that interest them. However, it doesn't suffice to settle the continuum problem!

So you see, the continuum refuses to be tamed!

And now we'll see how the real numbers, annoyed at being "defined" by Cantor and Dedekind, got their revenge in the century after Cantor, the 20th century.

BOREL'S AMAZING KNOW-IT-ALL REAL NUMBER

The first intimation that there might be something wrong, something terribly wrong, with the notion of a real number comes from a small paper published by Émile Borel in 1927.

Borel pointed out that if you really believe in the notion of a real number as an infinite sequence of digits 3.1415926 ... , then you could put all of human knowledge into a single real number. Well, that's not too difficult to do, that's only a finite amount of information. You just take your favorite encyclopedia, for example, the *Encyclopaedia Britannica,* which I used to use when I was in high-school—we had a nice library at the Bronx High School of Science—and you digitize it, you convert it into binary, and you use that binary as the base-two expansion of a real number in the unit interval between zero and one!

So that's pretty straightforward, especially now that most information, including books, is prepared in digital form before being printed.

But what's more amazing is that there's nothing to stop us from putting an infinite amount of information into a real number. In fact, there's a single real number, I'll call it Borel's number, since he imagined it, in 1927, that can serve as an oracle and answer any yes/no question that we could ever pose to it. How? Well, you just number all the possible questions, and then the Nth digit or Nth bit of Borel's number tells you whether the answer is yes or no!

If you could come up with a list of all possible yes/no questions and only valid yes/no questions, then Borel's number could give us the answer in its binary digits. But it's hard to do that. It's much easier to simply list all possible texts in the English language (and Borel did it using the French language), all possible finite strings of characters that you can form using the English alphabet, including a blank for use between words. You start with all the one-character strings, then all the two-character strings, etc. And you number them all like that . . .

Then you can use the *N*th digit of Borel's number to tell you whether the *N*th string of characters is a valid text in English, then whether it's a yes/no question, then whether it has an answer, then whether the answer is yes or no. For example, "Is the answer to this question 'No'?" looks like a valid yes/no question, but in fact has no answer.

So we can use *N*th digit 0 to mean bad English, 1 to mean not a yes/no question, 2 to mean unanswerable, and 3 and 4 to mean "yes" and "no" are the answers, respectively. Then 0 will be the most common digit, then 1, then there'll be about as many 3's as 4's, and, I expect, a smattering of 2's.

Now Borel raises the extremely troubling question, "Why should we believe in this real number that answers every possible yes/no question?" And his answer is that he doesn't see any reason to believe in it, none at all! According to Borel, this number is merely a mathematical fantasy, a joke, a *reductio ad absurdum* of the concept of a real number!

You see, some mathematicians have what's called a "constructive" attitude. This means that they only believe in mathematical objects that can be constructed, that, given enough time, in theory one could actually calculate. They think that there ought to be some way to **calculate** a real number, to calculate it digit by digit, otherwise in what sense can it be said to have some kind of mathematical existence?

And this is precisely the question discussed by Alan Turing in his famous 1936 paper that invented the computer as a mathematical concept. He showed that there were lots and lots of computable real numbers. That's the positive part of his paper. The negative part is that he also showed that there were lots and lots of uncomputable real numbers. And that gives us another philosophical proof that there are transcendental numbers, because it turns out that all algebraic reals are in fact computable.

TURING: UNCOMPUTABLE REALS ARE TRANSCENDENTAL

Turing's argument is very simple, very Cantorian in flavor. First he invents a computer (on paper, as a mathematical idea, a model computer). Then he points out that the set of all possible computer programs is a countable set, just like the set of all possible English texts. Therefore the set of all possible computable real numbers must also be countable. But the set of all reals is uncountable, it has the power of the continuum. Therefore the set of all uncomputable reals is also uncountable and has the power of the continuum. Therefore most reals are uncomputable, infinitely more than are computable.

That's remarkably simple, if you believe in the idea of a general-purpose digital computer. Now we are all very familiar with that idea. Turing's paper is long precisely because that was not at all the case in 1936. So he had to work out a simple computer on paper and argue that it could compute anything that can ever be computed, before giving the above argument that most real numbers will then be uncomputable, in the sense that there cannot be a program for computing them digit by digit forever.

The other difficult thing is to work out in detail precisely why algebraic reals can be computed digit by digit. Well, it's sort of intuitively obvious that this has to be the case; after all, what could possibly go wrong?! In fact, this is now well-known technology using something called Sturm sequences, that's the slickest way to do this; I'm sure that it comes built into *Mathematica* and *Maple,* two symbolic computing software packages. So you can use these software packages to calculate as many digits as you want. And you need to be able to calculate hundreds of digits in order to do research the way described by Jonathan Borwein and David Bailey in their book *Mathematics by Experiment.*

But in his 1936 paper Turing mentions a way to calculate algebraic reals that will work for a lot of them, and since it's a nice idea, I thought I'd tell you about it. It's a technique for root-solving by successive interval halving.

Let's write the algebraic equation that determines an individual algebraic real r that we are interested in as $\phi(x) = 0$; $\phi(x)$ is a polynomial in x. So $\phi(r) = 0$, and let's suppose we know two rational numbers α, β such that $\alpha < r < \beta$ and $\phi(\alpha) < \phi(r) < \phi(\beta)$ and we also know that there is no other root of the equation $\phi(x) = 0$ in that interval. So the signs of $\phi(\alpha)$ and $\phi(\beta)$ have to be different, neither of them is zero, and precisely one of them is greater than zero and one of them is less than zero, that's key. Because if ϕ changes from positive to negative it must pass through zero somewhere in between.

Then you just bisect this interval $[\alpha,\beta]$. You look at the midpoint $(\alpha + \beta)/2$, which is also a rational number, and you plug that into ϕ and you see whether or not $\phi((\alpha+\beta)/2)$ is equal to zero, less than zero, or greater than zero. It's easy to see which, since you're only dealing with rational numbers, not with real numbers, which have an infinite number of digits.

Then if ϕ of the midpoint gives zero, we have found r and we're finished. If not, we choose the left half or the right half of our original interval in such a way that the sign of ϕ at both ends is different, and this new interval replaces our original interval, r must be there, and we keep on going like that forever. And that gives us better and better approximations to the algebraic number r, which is what we wanted to show was possible, because at each stage the interval containing r is half the size it was before.

And this will work if r is what is called a "simple" root of its defining equation $\phi(r) = 0$, because in that case the curve for $\phi(x)$ will in fact cross zero at $x = r$. But if r is what is called a "multiple" root, then the curve may just graze zero, not cross it, and the Sturm sequence approach is the slickest way to proceed.

Now let's stand back and take a look at Turing's proof that there are transcendental reals. On the one hand, it's philosophical like Cantor's proof; on the other hand, it is some work to verify in detail that all algebraic reals are computable, although to me that seems obvious in some sense that I would be hard-pressed to justify/explain.

At any rate, now I'd like to take another big step, and show you that there are uncomputable reals in a very different way from the way that Turing did it, which is very much in the spirit of Cantor. Instead I'd like to use probabilistic ideas, ideas from what's called measure theory, which was developed by Lebesgue, Borel, and Hausdorff, among others, and which immediately shows that there are uncomputable reals in a totally un-Cantorian manner.

REALS ARE UNCOMPUTABLE WITH PROBABILITY ONE!

I got this idea from reading Courant and Robbins, *What is Mathematics?,* where they give a measure-theoretic proof that the reals are non-denumerable (more numerous than the integers).

Let's look at all the reals in the unit interval between zero and one. The total length of that interval is of course exactly one. But it turns out that all of the computable reals in it can be covered with intervals having total length exactly ϵ, and we can make ϵ as small as we want. How can we do that?

Well, remember that Turing points out that all the possible computer programs can be put in a list and numbered one by one, so there's a first program, a second program, and so forth and so on . . . Some of these programs don't compute computable reals digit by digit; let's just forget about them and focus on the others. So there's a first computable real, a second computable real, etc. And you just take the first computable real and cover it with an interval of size $\epsilon/2$, and you take the second computable real and you cover it with an interval of size $\epsilon/4$, and you keep going that way, halving the size of the covering interval each time. So the total size of all the covering intervals is going to be exactly

$$\frac{\epsilon}{2} + \frac{\epsilon}{4} + \frac{\epsilon}{8} + \frac{\epsilon}{16} + \frac{\epsilon}{32} + \ldots = \epsilon$$

which can be made as small as you like.

And it doesn't matter if some of these covering intervals fall partially outside of the unit interval, that doesn't change anything.

So all the computable reals can be covered this way, using an arbitrarily small part ϵ of the unit interval, which has length exactly equal to one.

So if you close your eyes and pick a real number from the unit interval at random, in such a way that any one of them is equally likely, the probability is zero that you get a computable real. And that's also the case if you get the successive binary digits of your real number using independent tosses of a fair coin. It's possible that you get a computable real, but it's infinitely unlikely. So with probability one you get an uncomputable real, and that has also got to be a transcendental number, what do you think of that!

Liouville, Hermite and Lindemann worked so hard to exhibit individual transcendentals, and now we can do it, almost certainly, by just picking a real number out of a hat! That's progress for you!

So let's suppose that you do that and get a specific uncomputable real that I'm going to call $R*$. What if you try to prove what some of its bits are when you write $R*$ in base-two binary?

Well, we've got a problem if we try to do that . . .

A FAS CANNOT DETERMINE INFINITELY MANY BITS OF AN UNCOMPUTABLE REAL

The problem is that if we are using a Hilbert/Turing/Post FAS, as we saw in Chapter Two there has got to be an algorithm for computably enumerating all the theorems, and so if we can prove what all the bits are, then we can just go ahead and calculate $R*$ bit by bit, which is impossible. To do that, you would just go through all the theorems one by one until you find the value of any particular bit of $R*$.

In fact, the FAS has got to fail infinitely many times to determine a bit of $R*$, otherwise we could just keep a little (finite) table on the side telling us which bits the FAS misses, and what their values are, and again we'd be able to compute $R*$, by combining the table with the FAS, which is impossible.

So our study of transcendentals now has a new ingredient, which is that we're using probabilistic methods. This is how I get a specific uncomputable real $R*$, not the way that Turing originally did it, which was using Cantor's diagonal method.

And now let's really start using probabilistic methods, let's start talking about algorithmically incompressible, irreducible random real numbers. What do I mean by that?

RANDOM REALS ARE UNCOMPUTABLE AND HAVE PROBABILITY ONE

Well, a random real is a real number with the property that its bits are irreducible, incompressible information, as much as is possible. Technically, the way you guarantee that is to require that the smallest self-delimiting binary program that calculates the first N bits of the binary expansion of R is always greater than $N - c$ bits long, for all N, where c is a constant that depends on R but not on N.

What's a "Random" Real Number R?

There's a constant c such that
H(the first N bits of R) $> N - c$ for all N.

The program-size complexity of the first N bits of R can never drop too far below N.

But I don't want to get into the details. The general idea is that you just require that the program-size complexity that we defined in Chapter Three of the first N bits of R has got to be as large as possible, as long as most reals can satisfy the lower bound that you put on this complexity. That is to say, you demand that the complexity be as high as possible, as long as the reals that can satisfy this requirement continue to have probability one, in other words, as long as the probability that a real fails to have complexity this high is zero.

So this is just a simple application of the ideas that we discussed in Chapter Three.

So of course this random incompressible real won't be computable, for computable reals only have a finite amount of information. But what if we try to use a particular FAS to determine particular bits of a particular random real $R*$? Now what?

A FAS CAN DETERMINE ONLY FINITELY MANY BITS OF A RANDOM REAL!

Well, it turns out that things have gotten worse, much worse than before. Now we can only determine finitely many bits of that random real $R*$.

Why?

Well, because if we could determine infinitely many bits, then those bits aren't really there in the binary expansion of $R*$, we get them for free, you see, by generating all the theorems in the FAS. You just see how many bits there are in the program for generating all the theorems, you see how complex the FAS is. Then you use the FAS to determine that many bits of $R*$, and a few more (just a little more than the c in the definition of the random real $R*$). And then you fill in the holes up to the last bit that you got using the FAS. (This can actually be done in a self-delimiting manner, because we already know exactly how many bits we need, we know that in advance.) And so that gives you a lot of bits from the beginning of the binary expansion of $R*$, but the program for doing it that I described is just a little bit too small, its size is substantially smaller than the number of bits of $R*$ that it got us, which contradicts the definition of a random real.

So random reals are bad news from the point of view of what can be proved. Most of their bits are unknowable; a given FAS can only determine about as many bits of the random real as the number of bits needed to generate the set of all its theorems. In other words, the bits of a random real, any finite set of them, cannot be compressed into a FAS with a smaller number of bits.

So by using random reals we get a much worse incompleteness result than by merely using uncomputable reals. You get at most a finite number of bits using any FAS. In fact, essentially the only way to prove what a bit of a particular random real is using a FAS is if you put that information directly into the axioms!

The bits of a random real are maximally unknowable!

So is there any reason to believe in such a random real? Plus I need to pick out a specific one, otherwise this incompleteness result isn't very interesting. It just says, pick the real $R*$ out of a hat, and then any particular FAS can prove what at most a finite number of bits are. But there is no way to even refer to that specific real number $R*$ within the FAS: it doesn't have a name!

Well, we'll solve that problem in the next chapter by picking out one random real, the halting probability Ω.

Meanwhile, we've realised that naming individual real numbers can be a problem. In fact, most of them **can't even be named.**

REALS ARE UN-NAMEABLE WITH PROBABILITY ONE!

The proof is just like the one that computable reals have probability zero. The set of all names for real numbers, if you fix your formal language or FAS, is just a countable infinity of names: because there's a first name, a second name, etc.

So you can cover them all using intervals that get smaller and smaller, and the total size of all the covering intervals is going to be exactly as before

$$\frac{\epsilon}{2} + \frac{\epsilon}{4} + \frac{\epsilon}{8} + \frac{\epsilon}{16} + \frac{\epsilon}{32} + \dots = \epsilon$$

which, as before, can be made as small as you like.

So, with probability one, a specific real number chosen at random cannot even be named uniquely, we can't specify it somehow, constructively or not, we can't define it or even refer to it!

So why should we believe that such an un-nameable real even exists?!

I claim that this makes incompleteness obvious: a FAS cannot even name all reals!

Instantaneous proof! We did it in three little paragraphs! Of course, this is a somewhat different kind of incompleteness than the one that Gödel originally exhibited.

People were very impressed by the technical difficulty of Gödel's proof. It had to be that difficult, because it involves constructing an assertion about whole numbers that can't be proved within a particular, popular FAS (called Peano arithmetic). But if you change whole numbers to real numbers, and if you talk about what you can name, rather than about what you can prove, then incompleteness is, as we've just seen, immediate!

So why climb the north face of the Eiger, when you can take a helicopter and have a picnic lunch on the summit in the noonday sun? Of course, there are actually plenty of reasons to climb that north face. I was recently proud to shake hands with someone who's tried to do it several times.

But in my opinion, this **is** the best proof of incompleteness! As Pólya says in *How to Solve It*, after you solve a problem, if you're a future mathematician, that's just the beginning. You should look back, reflect on what you did, and how you did it, and what the alternatives were. What other possibilities are there? How general is the method that you used? What was the key idea? What else is it good for? And can you do it without any computation, or can you see it at a glance?

This way, the way we just did it here, yes, you can certainly see incompleteness at a glance!

The only problem is that most people are not going to be too impressed by this particular kind of incompleteness, because it seems too darn philosophical. Gödel made it much more down to earth by talking about the positive integers, instead of exploiting problematical aspects of the reals.

And in spite of what mathematicians may brag, they have always had a slightly queasy attitude about reals, it's only with whole numbers that they feel really and absolutely confident.

And in the next chapter I'll solve that, at least for Ω. I'll take my paradoxical real, Ω, and dress it up as a diophantine problem, which just talks about whole numbers. This shows that even though Ω is a real number, you've got to take it seriously!

CHAPTER SUMMARY

In summary: **Why should I believe in a real number if I can't calculate it, if I can't prove what its bits are, and if I can't even refer to it? And each of these things happens with probability one!**

Against Real Numbers!

$$\mathbf{Prob}\{\text{algebraic reals}\} = \mathbf{Prob}\{\text{computable reals}\} = \mathbf{Prob}\{\text{nameable reals}\} = 0$$

$$\mathbf{Prob}\{\text{transcendental reals}\} = \mathbf{Prob}\{\text{uncomputable reals}\} = \mathbf{Prob}\{\text{random reals}\} = \mathbf{Prob}\{\text{un-nameable reals}\} = 1$$

We started this chapter with $\sqrt{2}$, which the ancient Greeks referred to as "unutterable," and we ended it by showing that with probability one there is no way to even name or to specify or define or refer to, no matter how non-constructively, individual real numbers. We have come full circle, from the unutterable to the un-nameable! There is no escape, these issues will not go away!

In the previous chapter we saw **physical arguments** against real numbers. In this chapter we've seen that reals are also problematic from a **mathematical** point of view, mainly because they contain an **infinite** amount of information, and infinity is something we can imagine but rarely touch. So I view these two chapters as validating the discrete, digital information approach of AIT, which does not apply comfortably in a physical or mathematical world made up out of real numbers. And I feel that this gives us the right to go ahead and see what looking at the size of computer programs can buy us, now that we feel reasonably comfortable with this new digital, discrete viewpoint, now that we've examined the philosophical underpinnings and the tacit assumptions that allowed us to posit this new concept.

In the next chapter I'll finally settle my two debts to you, dear reader, by proving that Turing's halting problem cannot be solved—my way, not the way that Turing originally did. And I'll pick out an individual random real, my Ω number, which as we saw in this chapter must have the property that any FAS can determine at most finitely many bits of its base-two binary expansion. And we'll discuss what the heck it all means, what it says about how we should do mathematics . . .

Six

COMPLEXITY, RANDOMNESS & INCOMPLETENESS

In Chapter Two I showed you Turing's approach to incompleteness. Now let me show you how I do it . . .

I am very proud of my two incompleteness results in this chapter! These are the jewels in the AIT crown, the best (or worst) incompleteness results, the most shocking ones, the most devastating ones, the most enlightening ones, that I've been able to come up with! Plus they are a consequence of the digital philosophy viewpoint that goes back to Leibniz and that I described in Chapter Three. That's why these results are so astonishingly different from the classical incompleteness results of Gödel (1931) and Turing (1936).

IRREDUCIBLE TRUTHS AND THE GREEK IDEAL OF REASON

I want to start by telling you about the very dangerous idea of "logical irreducibility" . . .

Mathematics:

axioms → **Computer** → theorems

We'll see in this chapter that the traditional notion of what math is about is all wrong: reduce things to axioms, compression. Nope, sometimes this doesn't work at all. The irreducible mathematical facts exhibited here in this chapter—the bits of Ω—**cannot** be derived from any principles simpler than they themselves are.

So the normal notion of the utility of proof fails for them—proof doesn't help at all in these cases. Simple axioms with complicated results is where proof helps. But here the axioms have to be as complicated as the result. So what's the point of using reasoning at all?!

Put another way: The normal notion of math is to look for structure and law in the world of math, for a theory. But theory implies compression, and here there cannot be any—there is no structure or law at all in this particular corner of the world of math.

And since there can be no compression, there can be no understanding of these mathematical facts!

In summary . . .

When Is Reasoning Useful?

"Axioms = Theorems" implies reasoning is useless!

"Axioms ≪ Theorems" implies compression & comprehension!

If the axioms are **exactly equal** in size to the body of interesting theorems, then reasoning was absolutely useless. But if the axioms are **much smaller** than the body of interesting theorems, then we have a substantial amount of compression, and so a substantial amount of understanding!

Hilbert, taking this tradition to its extreme, believed that a single FAS of finite complexity, a finite number of bits of information, must suffice to generate **all** of mathematical truth. He believed in a final theory of everything, at least for the world of pure math. The rich, infinite, imaginative, open-ended world of all of math, all of that compressed into a finite number of bits! What a magnificent compression that would have been! What a monument to the power of human reason!

COIN TOSSES, RANDOMNESS VS. REASON, TRUE FOR NO REASON, UNCONNECTED FACTS

And now, violently opposed to the Greek ideal of pure reason: Independent tosses of a fair coin, an idea from physics!

A "fair" coin means that it is as likely for heads to turn up as for tails. "Independent" means that the outcome of one coin toss doesn't influence the next outcome.

So each outcome of a coin toss is a unique, atomic fact that has no connection with any other fact: not with any previous outcome, not with any future outcome.

And knowing the outcome of the first million coin tosses, if we're dealing with independent tosses of a fair coin, gives us absolutely no help in predicting the very next outcome. Similarly, if we could know all the even outcomes (2nd coin toss, 4th toss, 6th toss), that would be no help in predicting any of the odd outcomes (1st toss, 3rd toss, 5th toss).

This idea of an infinite series of independent tosses of a fair coin may sound like a simple idea, a toy physical model, but it is a serious challenge, indeed a horrible nightmare, for any attempt to formulate a rational world view! Because each outcome is a fact that is true for **no reason,** that's true only by accident!

Rationalist Worldview

In the physical world, everything happens for a reason.

In the world of math, everything is true for a reason.

The universe is comprehensible, logical!

Kurt Gödel subscribed to this philosophical position.

So rationalists like Leibniz and Wolfram have always rejected physical randomness, or "contingent events," as Leibniz called them, because they cannot be understood using reason, they utterly refute the power of reason. Leibniz's solution to the problem is to claim that contingent events are also true for a reason, but in such cases there is in fact an infinite series of reasons, an infinite chain of cause and effect, that while utterly beyond the power of human comprehension is not at all beyond the power of comprehension of the divine mind. Wolfram's solution to the problem is to say that all of the **apparent** randomness that we see in the world is actually only **pseudo**randomness. It **looks like** randomness, but it is actually the result of simple laws, in the same way that the digits of $\pi = 3.1415926 \ldots$ **seem** to be random.

Nevertheless, at this point in time quantum mechanics demands real intrinsic randomness in the physical world, real unpredictability, and chaos theory even shows that a somewhat milder form of randomness is actually present in classical, deterministic physics, if you believe in infinite precision real numbers and in the power of acute sensitivity to initial conditions to quickly amplify random bits in initial conditions into the macroscopic domain . . .

The physicist Karl Svozil has the following interesting position on these questions. Svozil has classical, deterministic leanings and sympathizes with Einstein's assertion that "God doesn't play dice." Svozil admits that in its **current** state quantum theory contains randomness. But he thinks that this is only temporary, and that some new, deeper, hidden-variable theory will eventually restore determinacy and law to physics. On the other hand, Svozil believes that, as he puts it, Ω shows that there is **real** randomness in the (unreal) mental mindscape fantasy world of pure math!

A CONVERSATION ON RANDOMNESS AT CITY COLLEGE, NY, 1965

As background information for this story, I should start by telling you that a century ago Borel proposed a mathematical definition of a random real number, in fact, infinitely many variant definitions. He called such reals "normal" numbers. This is the same Borel who in 1927 invented the know-it-all-real that we discussed in Chapter Five. And in 1909 he was able to show that most real numbers must satisfy **all** of his variant definitions of normality. The probability is zero of failing to satisfy any of them.

What is Borel's definition of a normal real? Well, it's a real number with the property that every possible digit occurs with equal limiting frequency: in the long run, exactly 10% of the time. That's called "simple" 10-normal. And 10-normal *tout court* means that for each k, the base-ten expansion of the real has each of the 10^k possible blocks of k successive digits with exactly the same limiting relative frequency $1/10^k$. All the possible blocks of k successive digits are exactly equally likely to appear in the long run, and this is the case for **every** k, for $k = 1, 2, 3, \ldots$ Finally there is just plain "normal," which is a real that has this property when written in **any** base, not just base-ten. In other words, normal means that it's 2-normal, 3-normal, 4-normal, and so forth and so on.

So most reals are normal, with probability one (Borel, 1909). But what if we want to exhibit a **specific** normal number? And there seems to be no reason to doubt that π and e are normal, but no one, to this day, has the slightest idea how to prove this.

Okay, that's the background information on Borel normality. Now let's fast forward to 1965.

I'm an undergraduate at City College, just starting my second year. I was writing and revising my very first paper on randomness. My definition of randomness was not at all like Borel's. It was the much more demanding definition that I've discussed in this book, that's based on the idea of algorithmic incompressibility, on the idea of looking at the size of computer programs. In fact, it's actually based on a remark that Leibniz made in 1686, though I wasn't aware of that at the time. This is my paper that was published in 1966 and 1969 in the *ACM Journal.*

The Dean had excused me from attending classes so that I could prepare all this for publication, and the word quickly got around at City College that I was working on a new definition of randomness. And it turned out that there was a professor at City College, Richard Stoneham, whom I had never had in any of my courses, who was very interested in normal numbers.

We met in his office in wonderful old stone pseudo-gothic Shepard Hall, and I explained to him that I was working on a definition of randomness, one that would imply Borel normality, and that it was a definition that most reals would satisfy, with probability one.

He explained that he was interested in proving that **specific** numbers like π or e were normal. He wanted to show that some well-known mathematical object already contained randomness, Borel normality, not some new kind of randomness. I countered that no number like π or e could satisfy my much more demanding definition of randomness, because they were computable reals and therefore compressible.

He gave me some of his papers. One paper was computational and was on the distribution of the different digits in the decimal expansions of π and e. They seemed to be normal . . . And another paper was theoretical, on digits in rational numbers, in periodic decimal expansions. Stoneham was able to show that for some rationals m/n the digits were sort of equi-distributed, subject to conditions on m and n that I no longer recall.

That was it, that was the only time that we met.

And the years went by, and I came up with the halting probability Ω. And I never heard of Stoneham again, until I learned in Borwein and Bailey's chapter on normal numbers in *Mathematics by Experiment* that Stoneham had actually managed to do it! Amazingly enough, we had **both** achieved our goals!

Bailey and Crandall, during the course of their own work in 2003 that produced a much stronger result, had discovered that 30 years before, Stoneham, now deceased (that's information from Wolfram's book), had succeeded in finding what as far as I know was the first "natural" example of a normal number.

Looking like a replay of Liouville, Stoneham had not been able to show that π or e were normal. But he had been able to show that the sum of a natural-looking infinite series was 2-normal, that is, normal for blocks of bits of every possible size in base-2!

He and I **both** found what we were looking for! How delightful!

And in this chapter I'll tell you how we did it. But first, as a warm-up exercise, to get in the right mood, I want to prove that Turing's halting problem is unsolvable. We should do that before looking at my halting probability Ω.

And to do that, I want to show that you can't prove that a program is elegant, except finitely often. Believe it or not, the idea for that proof actually comes from **Émile Borel,** the same Borel as before. Although I wasn't aware of this when I originally found the proof on my own . . .

So let me start by telling you Borel's beautiful idea.

BOREL'S UNDEFINABILITY-OF-RANDOMNESS PARADOX

I have a very clear and distinct recollection of reading the idea that I'm about to explain to you, in an English translation of a book by Borel on the basic ideas of probability theory. Unfortunately, I have never been able to discover which book it was, nor to find again the following discussion by Borel, about a paradox regarding any attempt to give a definitive notion of randomness.

So it is possible that this is a false memory, perhaps a dream that I once had, which sometimes seem real to me, or a "reprocessed" memory that I have somehow fabricated over the years.

However, Borel was quite prolific, and was very interested in these issues, so this is probably somewhere in his oeuvre! If you find it, please let me know!

That caution aside, let me share with you my recollection of Borel's discussion of a problem that must inevitably arise with **any** attempt to define the notion of randomness.

Let's say that somehow you can distinguish between whole numbers whose decimal digits form a particularly random sequence of digits, and those that don't.

Now think about the first *N*-digit number that satisfies your definition of randomness. But this particular number is rather atypical, because it happens to be precisely the first *N*-digit number that has a particular property!

The problem is that random means "typical, doesn't stand out from the crowd, no distinguishing features." But if you can define randomness, then the property of being random becomes just one more feature that you can use to show that certain numbers are atypical and stand out from the crowd!

So in this way you get a hierarchy of notions of randomness, the one you started out with, then the next one, then one derived from that, and so forth and so on . . . And each of these is derived using the previous definition of randomness as one more characteristic just like any other that can be used to classify numbers!

Borel's conclusion is that there can be no one definitive definition of randomness. You can't define an all-inclusive notion of randomness. Randomness is a slippery concept, there's something paradoxical about it, it's hard to grasp. It's all a matter of deciding how much we want to demand. You have to decide on a cut-off, you have to say "enough," let's take **that** to be random.

Let me try to explain this by using some images that are definitely not in my possibly false Borel recollection.

The moment that you fix in your mind a notion of randomness, that very mental act invalidates that notion and creates a new more demanding notion of randomness . . . So fixing randomness in your mind is like trying to stare at something without blinking and without moving your eyes. If you do that, the scene starts to disappear in pieces from your visual field. To see something, you have to keep moving your eyes, changing your focus of attention . . .

The harder you stare at randomness, the less you see it! It's sort of like trying to see faint objects in a telescope at night, which you do by not staring directly at them, but instead looking aside, where the resolution of the retina is lower and the color sensitivity is lower, but you can see much fainter objects . . .

This discussion that I shall attribute to Borel in fact contains the germ of the idea of my proof that I'll now give that you can't prove that a computer program is "elegant," that is to say, it's the smallest program that produces the output that it does. More precisely, a program is elegant if no program smaller than it produces the same output. You see, there may be a tie, there may be several different programs that have exactly the same minimum-possible size and produce the same output.

Viewed as a theory, as discussed in Chapter Three, an elegant program is the optimal compression of its output, it's the simplest scientific theory for that output, considered as experimental data.

So if the output of this computer program is the entire universe, then an elegant program for it would be the optimal TOE, the optimal Theory of Everything, the one with no redundant elements, the best TOE, the one, Leibniz would say, that a perfect God would use to produce that particular universe.

WHY CAN'T YOU PROVE THAT A PROGRAM IS "ELEGANT"?

So let's consider a Hilbert/Turing/Post formal axiomatic system FAS, which (as discussed in Chapter Two) for us is just a program for generating all the theorems. And we'll assume that this program is the smallest possible one that produces that particular set of theorems. So the size of this program is precisely the program-size complexity of that theory. There's absolutely no redundancy!

Okay, so that's how we can measure the power of FAS by how many bits of information it contains. As you'll see now this really works, it really gives us new insight.

Building on Borel's paradox discussed in the previous section, consider this computer program . . .

> **Paradoxical Program P:**
>
> The output of P is the same as the output of the first provably elegant program Q that you encounter (as you generate all the theorems of your chosen FAS) that is larger than P.

Why are we building on Borel's paradox? Because the idea is that if we could prove that a program is elegant, then that would enable us to find a smaller program that produces the same output, contradiction! Borel's paradox is that if we could define randomness, then that would enable us to pick out a random number that isn't at all random, contradiction. So I view these two proofs, Borel's informal paradox, and this actual theorem, or, more precisely, meta-theorem, as being in the same spirit.

In other words, P generates all the theorems of the FAS until it finds a proof that a particular program Q is elegant, and what's more the size of Q has got to be larger than the size of P. If P can find such a Q, then it runs Q and produces Q's output as its own output.

Contradiction, because P is too small to produce the same output as Q, since P is smaller than Q and Q is supposed to be elegant (assuming that all theorems proved in the FAS are correct, are true)! The only way to avoid this contradiction is that P never finds Q, because no program Q that's larger than P is ever shown to be elegant in this FAS.

So using this FAS, you can't prove that a program Q is elegant if it's larger than P is. So in this FAS, you can't prove that more than finitely many specific programs are elegant. (So the halting problem is unsolvable, as we'll see in the next section.)

And P is just a fixed number of bits larger than the FAS that we're using. Mostly P contains the instructions for generating all the theorems, that's the variable part, plus a little more, a fixed part, to filter them and carry out our proof as above.

So the basic idea is that you can't prove that a program is elegant if it's larger than the size of the program for generating all the theorems in your FAS. In other words, if the program is larger than the program-size complexity of your FAS, then you can't prove that that program is elegant! No way!

You see how useful it is to be able to measure the complexity of a FAS, to be able to measure how many bits of information it contains?

Tacit assumption: In this discussion we've assumed the soundness of the FAS, which is to say, we've assumed that all the theorems that it proves are true.

Caution: We don't really have complete irreducibility yet, because the different cases "*P* is elegant," "*Q* is elegant," "*R* is elegant" are **not** independent of one another. In fact, you can determine all elegant programs less than *N* bits in size **from the same** *N*-bit axiom, one that tells you which program with less than *N* bits takes longest to halt. Can you see how to do this?

But before we solve this problem and achieve total irreducibility with the bits of the halting probability Ω, let's pause and deduce a useful corollary.

BACK TO TURING'S HALTING PROBLEM

Here's an immediate consequence, it's a corollary, of the fact that we've just established that you can't mechanically find more than finitely many elegant programs:

There is no algorithm to solve the halting problem, to decide whether or not a given program ever halts. Proof by *reductio ad absurdum:* Because if there **were such an algorithm,** we could use it to find all the elegant programs. You'd do this by checking in turn every program to see if it halts, and then running the ones that halt to see what they produce, and then keeping only the first program you find that produces a given output. If you look at all the programs in size order, this will give you precisely all the elegant programs (barring ties, which are unimportant and we can forget about).

In fact, we've actually just shown that if our halting problem algorithm is N bits in size, then there's got to be a program that never halts, and that's at most just a few bits larger than N bits in size, but we can't decide that this program never halts using our N-bit halting problem algorithm. (This assumes that the halting problem algorithm prefers never to give an answer rather than to give the wrong answer. Such an algorithm can be reinterpreted as a FAS.)

So we've just proved Turing's famous 1936 result completely differently from the way that he originally proved it, using the idea of program size, of algorithmic information, of software complexity. Of all the proofs that I've found of Turing's result, this one is my favorite.

And now I must confess something else. Which is that I don't think that you can really understand a mathematical result until you find your own proof. Reading somebody else's proof is not as good as finding your own proof. In fact, one fine mathematician that I know, Robert Solovay, never let me explain a proof to him. He would always insist on just being told the statement of the result, and then he would think it through on his own! I was very impressed!

This is the most straightforward, the most direct, the most basic proof of Turing's result that I've been able to find. I've tried to get to the heart of the matter, to remove all the clutter, all the peripheral details that get in the way of understanding.

And this also means that this missing piece in our Chapter Two discussion of Hilbert's 10th problem has now been filled in.

Why is the halting problem interesting? Well, because in Chapter Two we showed that if the halting problem is unsolvable, then Hilbert's 10th problem cannot be solved, and there is no algorithm to decide whether a diophantine equation has a solution or not. In fact, Turing's halting problem is equivalent to Hilbert's 10th problem, in the sense that a solution to either problem would automatically provide/entail a solution for the other one.

THE HALTING PROBABILITY Ω AND SELF-DELIMITING PROGRAMS

Now we're really going to get irreducible mathematical facts, mathematical facts that "are true for no reason," and which simulate in pure math, as much as is possible, independent tosses of a fair coin: It's the bits of the base-two expansion of the halting probability Ω. The beautiful, illuminating fact that you can't prove that a program is elegant was just a warm-up exercise!

Instead of looking at **one** program, like Turing does, and asking whether or not it halts, let's put **all possible** programs in a bag, shake it up, close our eyes, and pick out a program. What's the probability that this program that we've just chosen at random will eventually halt? Let's express that probability as an infinite precision binary real between zero and one. And *voila!*, its bits are our independent mathematical facts.

The Halting Probability Ω

You run a program chosen by chance on a fixed computer.
Each time the computer requests the next bit of the program,
flip a coin to generate it, using independent tosses of a fair coin.
The computer must decide **by itself** when to stop reading
the program. This forces the program to be self-delimiting
binary information.

You sum for each program that halts
the probability of getting precisely
that program by chance:

$$\Omega = \sum_{\text{program } p \text{ halts}} 2^{-(\text{size in bits of } p)}$$

Each k-bit self-delimiting program p that halts
contributes $1/2^k$ to the value of Ω.

The self-delimiting program proviso is crucial: Otherwise the halting probability has to be defined for programs of **each particular size,** but it cannot be defined over **all** programs of **arbitrary size.**

To make Ω seem more real, let me point out that you can compute it in the limit from below:

Nth Approximation to Ω

Run each program up to N bits in size for N seconds.

Then each k-bit program you discover that halts contributes $1/2^k$ to this approximate value for Ω.

These approximate values get bigger and bigger (slowly!) and they approach Ω more and more closely, from below.

1st approx. ≤ 2nd approx. ≤ 3rd approx. ≤ ... ≤ Ω

This process is written in LISP in my book *The Limits of Mathematics.* The LISP function that gives this approximate value for Ω as a function of N is about half a page of code using the special LISP dialect that I present in that book.

However, this process converges very, very slowly to Ω. In fact, you can never know how close you are, which makes this a rather weak kind of convergence.

Normally, for approximate values of a real number to be useful, you have to know how close they are to what they're approximating, you have to know what's called "the rate of convergence," or have what's called "a computable regulator of convergence." But here we don't have any of that, just a sequence of rational numbers that very slowly creeps closer and closer to Ω, but without enabling us to ever know precisely how close we are to Ω at a given point in this unending computation.

In spite of all of this, these approximations are extremely useful: In the next section, we'll use them in order to show that Ω is an algorithmically "random" or "irreducible" real number. And later in this chapter, we'll use them to construct diophantine equations for the bits of Ω.

Ω AS AN ORACLE FOR THE HALTING PROBLEM

Why are the bits of Ω irreducible mathematical facts? Well, it's because we can use the first N bits of Ω to settle the halting problem for all programs up to N bits in size. That's N bits of information, and the first N bits of Ω are therefore an irredundant representation of this information.

How much information is there in the first N bits of Ω?

Given the first N bits of Ω, get better and better approximations for Ω as indicated in the previous section, until the first N bits of the approximate value are correct.

At that point you've seen every program up to N bits long that ever halts. Output something not included in any of the output produced by all these programs that halt. It cannot have been produced using any program having less than or equal to N bits.

Therefore the first N bits of Ω cannot be produced with any program having substantially less than N bits, and Ω satisfies the definition of a "random" or "irreducible" real number given in Chapter Five:

$$H(\text{the first } N \text{ bits of } \Omega) \; > \; N - c$$

This process is written in LISP in my book *The Limits of Mathematics.* The LISP function that produces something with complexity greater than N bits if it is given any program that calculates the first N

bits of Ω is about a page of code using the special LISP dialect that I present in that book. The size in bits of this one-page LISP function is precisely the value of that constant c with the property that H(the first N bits of Ω) is greater than $N - c$ for all N. So, the program-size complexity of the first N bits of Ω never drops too far below N.

Now that we know that Ω is an algorithmically "random" or "irreducible" real number, the argument that a FAS can determine only finitely many bits of such a number given in Chapter Five immediately applies to Ω. The basic idea is that if K bits of Ω could be "compressed" into a substantially less than K-bit FAS, then Ω wouldn't really be irreducible. In fact, using the argument given in Chapter Five, we can say exactly how many bits of Ω a given FAS can determine. Here's the final result . . .

A FAS can only determine as many bits of Ω as its complexity

As we showed in Chapter Five, there is (another) constant c such that a formal axiomatic system FAS with program-size complexity H(FAS) can never determine more than H(FAS) + c bits of the value for Ω.

These are theorems of the form "The 39th bit of Ω is 0" or "The 64th bit of Ω is 1."

(This assumes that the FAS only enables you to prove such theorems if they are true.)

This is **an extremely strong incompleteness result,** it's the very best I can do, because it says that essentially the only way to determine bits of Ω is to put that information directly into the axioms of our FAS, without using any reasoning at all, only, so to speak, table look-up to determine these finite sets of bits.

In other words, the bits of Ω are logically irreducible, they cannot be obtained from axioms simpler than they are. Finally! We've found a way to simulate independent tosses of a fair coin, we've found "atomic" mathematical facts, an infinite series of math facts that have no connection with each other and that are, so to speak, "true for no reason" **(no reason simpler than they are).**

Also, this result can be interpreted informally as saying that math is random, or more precisely, contains randomness, namely the bits of Ω. What a dramatic conclusion! But a number of serious caveats are in order!

Math isn't random in the sense of being arbitrary, not at all—it most definitely is not the case that $2 + 2$ is occasionally equal to 5 instead of 4! But math does contain **irreducible information,** of which Ω is the prime example.

To say that Ω is random may be a little confusing. It's a specific well-determined real number, and technically it satisfies the definition of what I've been calling a "random real." But math often uses familiar words in unfamiliar ways. Perhaps a better way to put it is to say that Ω is algorithmically incompressible. Actually, I much prefer the word "irreducible"; I'm coming to use it more and more, although for historical reasons the word "random" is unavoidable.

So perhaps it's best to avoid misunderstandings and to say that Ω is irreducible, which is true both algorithmically or computationally, and **logically,** by means of proofs. And *which happens to imply* that Ω has many of the characteristics of the typical outcome of a random process, in the physical sense of an unpredictable process amenable to statistical treatment.

For example, as we'll discuss in the next section, in base two, in Ω's infinite binary expansion, each of the 2^k possible k-bit blocks will appear with limiting relative frequency exactly $1/2^k$, and this is provably the case for the specific real number Ω, even though it's only true with probability one, but not with certainty, for the outcome of an infinite series of independent tosses of a fair coin. So perhaps, in retrospect, the choice of the word "random" wasn't so bad after all!

Also, a random real may be meaningless, or it may be extremely meaningful; my theory cannot distinguish between these two possibilities, it cannot say anything about that. If the real was produced using an independent toss of a fair coin for each bit, it'll be irreducible and it'll be meaningless. On the other hand, Ω is a random real with lots of meaning, since it contains a lot of information about the halting problem, and this information is stored in Ω in an irreducible fashion, with no redundancy. You see, once you compress out all the redundancy from anything meaningful, the result necessarily **looks** meaningless, even though it isn't, not at all, it's just jam-packed with meaning!

BOREL NORMAL NUMBERS AGAIN

In fact, it's not difficult to see that Ω is normal, that is to say, b-normal for any base b, not just 2-normal. And the lovely thing about this is that the definition of Ω as the halting probability doesn't seem to have anything to do with normality. Ω wasn't constructed especially with normality in mind; that just dropped out, for free, so to speak!

So for any base b and any fixed number k of base-b "digits," the limiting relative frequency of each of the b^k possible k-digit sequences in Ω's b-ary expansion will be exactly $1/b^k$. In the limit, they are all equally likely to appear . . .

How can you prove that this has got to be the case? Well, if this were **not** the case, then Ω's bits would be highly compressible, by a fixed multiplicative factor depending on just how unequal the relative frequencies happen to be. In other words, the bits of Ω could be compressed by a fixed percentage, which is a lot of compression . . . These are ideas that go back to a famous paper published by Claude Shannon in *The Bell System Technical Journal* in the 1940's, although the context in which we are working is rather different from his. Anyway that's where I got the idea, by reading Shannon. Shannon and I both worked for industrial labs: in his case, it was the phone company's, in my case, IBM's.

So that's how I succeeded in finding a normal number! It was because I wasn't interested in normality per se, but in deeper, philosophical issues, and normality just dropped out as an application of these ideas. Sort of similar to Turing's proof that transcendentals exist, because all algebraic numbers have to be computable . . .

And how did the City College professor Richard Stoneham succeed in **his** quest for randomness? Here it is, here is Stoneham's provably 2-normal number:

$$\frac{1}{(3 \times 2^{3})} + \frac{1}{(3^2 \times 2^{3^2})} + \frac{1}{(3^3 \times 2^{3^3})} + \cdots \frac{1}{(3^k \times 2^{3^k})} + \cdots$$

This is normal base 2 for blocks of all size.

And what is the more general result obtained by David Bailey and Richard Crandall in 2003?

$$\frac{1}{(c \times b^{c})} + \frac{1}{(c^2 \times b^{c^2})} + \frac{1}{(c^3 \times b^{c^3})} + \cdots \frac{1}{(c^k \times b^{c^k})} + \cdots$$

This is b-normal as long as b is greater than 1 and b and c have no common factors. For example, $b = 5$ and $c = 7$ will do; that gives you a 5-normal number. For the details, see the chapter on normal numbers, Chapter Four, in Borwein and Bailey, *Mathematics by Experiment.* And there's lots of other interesting stuff in that book, for example, amazing new ways to calculate π—which actually happens to be connected with these normality proofs! (See their Chapter Three.)

GETTING BITS OF Ω USING DIOPHANTINE EQUATIONS

In Chapter Five, I expressed **mathematical** skepticism about real numbers. And in Chapter Four, I expressed **physical** skepticism about real numbers. So why should we take the real number Ω seriously? Well, it's not just **any** real number, you can get it from a diophantine equation! In fact, you can do this in two rather different ways.

One approach, that I discovered in 1987, makes the number of solutions of an equation jump from finite to infinite in a way that mimics the bits of Ω. The other approach, discovered by Toby Ord and Tien D. Kieu in Australia in 2003, makes the number of solutions of the equation jump from even to odd in a way that mimics the bits of Ω. So take your choice; it can be done either way. There's a Northern and there's a Southern Hemisphere approach to this problem—and no doubt many other interesting ways to do it!

Remember Kronecker's credo that "God created the integers; all else is the work of man"? If you prefer, Ω isn't a real number at all, it's a fact about certain diophantine equations; it has to do only with whole numbers, with positive integers! So you cannot shrug away the fact that the bits of the halting probability Ω are irreducible mathematical truths, for this can be reinterpreted as a statement about diophantine equations.

Chaitin (1987):
Exponential Diophantine Equation #1

In this equation n is a **parameter,**
and $k, x, y, z, \ldots$ are the **unknowns:**

$$L(n, k, x, y, z, \ldots) \quad = \quad R(n, k, x, y, z, \ldots).$$

It has **infinitely many** positive-integer solutions
if the nth bit of Ω is a 1.

It has **only finitely many** positive-integer solutions
if the nth bit of Ω is a 0.

Ord, Kieu (2003):
Exponential Diophantine Equation #2

In this equation n is a **parameter,**
and $k, x, y, z, \ldots$ are the **unknowns:**

$$L(n, k, x, y, z, \ldots) \quad = \quad R(n, k, x, y, z, \ldots).$$

For any given value of the parameter n,
it only has finitely many positive-integer solutions.

For each particular value of n:

the number of solutions of this equation will be **odd**
if the nth bit of Ω is a 1, and

the number of solutions of this equation will be **even**
if the nth bit of Ω is a 0.

How do you construct these two diophantine equations? Well, the details get a little messy. The boxes below give you the general idea; they summarize what needs to be done. As you'll see, that computable sequence of approximate values of Ω that we discussed before plays a key role. It's also important, particularly for Ord and Kieu (2003), to recall that these approximate values are a **non-decreasing** sequence of rational numbers that get closer and closer to Ω but that always remain **less than** the value of Ω.

Chaitin (1987):
Exponential Diophantine Equation #1

Program(n,k) calculates the kth approximation to Ω, in the manner explained in a previous section. Then Program(n,k) looks at the nth bit of this approximate value for Ω. If this bit is a 1, then Program(n,k) immediately halts; otherwise it loops forever. So Program(n,k) halts if and only if (the nth bit in the kth approximation to Ω) is a 1.

As k gets larger and larger, the nth bit of the kth approximation to Ω will eventually settle down to the correct value. Therefore for all sufficiently large k:

Program(n,k) will halt if the nth bit of Ω is a 1,

and Program(n,k) will fail to halt if the nth bit of Ω is a 0.

Using all the work on Hilbert's 10th problem that we explained in Chapter Two, we immediately get an exponential diophantine equation

$$L(n, k, x, y, z, \ldots) \quad = \quad R(n, k, x, y, z, \ldots)$$

that has **exactly one** positive-integer solution if Program(n,k) eventually halts,

and that has **no** positive-integer solution if Program(n,k) never halts.

Therefore, fixing n and considering k to be an unknown, this exact same equation

$$L(n, k, x, y, z, \ldots) \quad = \quad R(n, k, x, y, z, \ldots)$$

has **infinitely many solutions** if the nth bit of Ω is a 1,

and it has **only finitely many** solutions if the nth bit of Ω is a 0.

Ord, Kieu (2003): Exponential Diophantine Equation #2

Program(n,k) halts if and only if $k > 0$ and
$2^n \times$ (jth approximation to Ω) $>$ k
for some $j = 1, 2, 3, \ldots$

So Program(n,k) halts if and only if $2^n \times \Omega > k > 0$.

Using all the work on Hilbert's 10th problem that we explained in Chapter Two, we immediately get an exponential diophantine equation

$$L(n, k, x, y, z, \ldots) = R(n, k, x, y, z, \ldots)$$

that has **exactly one** positive-integer solution if Program(n,k) eventually halts,

and that has **no** positive-integer solution if Program(n,k) never halts.

Let's fix n and ask for which k does this equation have a solution.

Answer: $L(n,k) = R(n,k)$ is solvable precisely for $k = 1, 2, 3, \ldots$, up to the integer part of $2^n \times \Omega$.

Therefore, $L(n) = R(n)$ has exactly integer part of $2^n \times \Omega$ solutions, which is the integer you get by shifting the binary expansion for Ω left n bits. And the right-most bit of the integer part of $2^n \times \Omega$ will be the nth bit of Ω.

Therefore, fixing n and considering k to be an unknown, this exact same equation

$$L(n, k, x, y, z, \ldots) = R(n, k, x, y, z, \ldots)$$

has **an odd number** of solutions if the nth bit of Ω is a 1,

and it has **an even number** of solutions if the nth bit of Ω is a 0.

WHY IS THE PARTICULAR RANDOM REAL Ω INTERESTING?

This is a good place to discuss a significant issue. In the previous chapter we pointed out that with probability one a real number is algorithmically irreducible. Algorithmically compressible reals have probability zero. So why is the **particular** random real Ω of any interest?! There are certainly plenty of them!

Well, for a number of reasons.

First of all, Ω links us to Turing's famous result; the halting problem is unsolvable and the halting probability is random! Algorithmic unsolvability in one case, algorithmic randomness or incompressibility in the other. Also, the first *N* bits of Ω give us a lot of information about particular, individual cases of the halting problem.

But the main reason that Ω is interesting is this: We have reached into the infinitely dark blackness of random reals and picked out **a single** random real! I wouldn't say that we can touch it, but we can certainly point straight at it. And it is important to make this obscurity tangible by exhibiting a specific example. After all, why should we believe that most things have a certain property if we can't exhibit a specific thing that has this property?

(Please note, however, that in the case of the un-nameable reals, which also have probability one, we're never ever going to be able to pick out an individual un-nameable real!)

Here's another way to put it: Ω is violently, maximally uncomputable, but it **almost** looks computable. It's just across the border between what we can deal with and things that transcend our abilities as mathematicians. So it serves to establish a sharp boundary, it draws a thin line in the sand that we dare not cross, that we **cannot** cross!

And that's also connected with the fact that we can compute better and better lower bounds on Ω, we just can't ever know how close we've gotten.

In other words, for an incompleteness result to be really shocking, the situation has got to be that just as we were about to reach out and touch something, we get our fingers slapped. We can never ever have it, even though it's attractively laid out on the dining room table, next to a lot of other inviting food! That's much more frustrating and much more interesting than being told that we can't have something or do something that never seemed concrete enough or down to earth enough that this ever looked like a legitimate possibility in the first place!

WHAT IS THE MORAL OF THE STORY?

So the world of mathematical truth has infinite complexity, even though any given FAS only has finite complexity. In fact, even the world of diophantine problems has infinite complexity, no finite FAS will do.

I therefore believe that we cannot stick with a single FAS, as Hilbert wanted, we've got to keep adding new axioms, new rules of inference, or some other kind of new mathematical information to the foundations of our theory. And where can we get new stuff that cannot be deduced from what we already know? Well, I'm not sure, but I think that it may come from the same place that physicists get their new equations: based on inspiration, imagination and on—in the case of math, computer, not laboratory—experiments.

So this is a "quasi-empirical" view of how to do mathematics, which is a term coined by Lakatos in an article in Thomas Tymoczko's interesting collection *New Directions in the Philosophy of Mathematics.* And this is closely connected with the idea of so-called "experimental mathematics," which uses computational evidence rather than conventional proof to "establish" new truths. This research methodology, whose benefits are argued for in a two-volume work by Borwein, Bailey and Girgensohn, may not only sometimes be **extremely convenient,** as they argue, but in fact it may sometimes even be **absolutely necessary** in order for mathematics to be able to progress in spite of the incompleteness phenomenon . . .

Okay, so that's **my** approach to incompleteness, and it's rather different from Gödel's and Turing's. The main idea is to measure the complexity or the information content of a FAS by the size in bits of the smallest program for generating all the theorems. Once you do that, everything else follows. From that initial insight, it's more or less straightforward, the developments are more or less systematic.

Yes, but as Pólya pointedly asks, "Can you see it all at a glance?" Yes, in fact I think that we can:

No mechanical process (rules of the game) can be really creative, because in a sense anything that ever comes out was already contained in your starting point. Does this mean that physical randomness, coin tossing, something non-mechanical, is the only possible source of creativity?! At least from this (enormously over-simplified!) point of view, it is.

AGAINST EGOTISM

The history of ideas offers many surprises. In this book we've seen that:

- Digital philosophy can be traced back to Leibniz, and digital physics can be traced back to Zeno.
- My definition of randomness via complexity goes back to Leibniz.
- My Ω number can be traced back to Borel's know-it-all real number. Borel's number is also an example of what Turing would later call an uncomputable real.
- The main idea of my proof that you can't prove that a program is elegant—in fact it's the basic idea of **all** my incompleteness results—can be traced back to Borel's undefinability-of-randomness paradox.

I think that these observations should serve as an antidote to the excessive egotism, competitiveness and the foolish fights over priority that poison science. No scientific idea has only one name on it; they are the joint production of the best minds in the human race, building on each other's insights over the course of history.

And the fact that these ideas can be traced that far back, that the threads are that long, doesn't weaken them in any way. On the contrary, it gives them even greater significance.

As my friend Jacob T. Schwartz once told me, the medieval cathedrals were the work of many hands, anonymous hands, and took lifetimes to build. And Schwartz delighted in quoting a celebrated doctor from that period, who said about a patient, "I treated him and God cured him!" I think that this is also the right attitude to have in science and mathematics.

Seven

CONCLUSION

As you have no doubt noticed, this is really a book on philosophy, not just a math book. And as Leibniz says in the quote at the beginning of the book, math and philosophy are inseparable.

Among the great philosophers, only Pythagoras and Leibniz were great mathematicians. Indeed, Pythagoras, though lost in the mists of time, is credited with inventing philosophy and mathematics and also inventing the **words** "philosophy" and "mathematics." As for Leibniz, he is what the Germans call a *Universalgenie*, a universal genius, interested in everything; and if he was interested in something, he always came up with an important new idea, he always made a good suggestion.

Since Leibniz, perhaps only Poincaré was a little like this. He was enough of a philosopher to come up with a version of relativity theory before Einstein did, and his popular books of essays were full of philosophical observations and are still in print.

And Leibniz used the word "transcendental," as in transcendental curves, numbers, and methods etc., deliberately, thinking of God's transcendence of all things human, which also inspired Cantor to develop his theory of infinite magnitudes. After all, math deals with the world of ideas, which transcends the real world. And for "God" you can understand the laws of the universe, as Einstein did, or the entire world, as Spinoza did, that doesn't change the message.

The topic of monotheism and polytheism is also germane here. I love the complexity and sophistication of South Indian vegetarian cuisine, and my home is decorated with Indian sculpture and fabrics—and much more. I admired Peter Brook's *Mahabharata,* which clearly brings out the immense philosophical depth of this epic.

But my intellectual personality is resolutely monotheistic. Why do I say that my personality is monotheistic? I mean it only in this rather abstract sense: that I am always searching for simple, unifying ideas, rather than glorying intellectually in "polytheistic" subjects like biology, in which there is a rich tapestry of extremely complicated facts that resists being reduced to a few simple ideas.

And let's look at Hilbert's FAS's. They failed miserably. Nevertheless formalism has been a brilliant success this past century, but not in math, not in philosophy, but as computer technology, as software, as programming languages. It works for machines, but not for us![8]

Just look at Bertrand Russell's three-volume magnum opus (with Whitehead) *Principia Mathematica,* which I keep in my office at IBM. One entire formula-filled volume to prove that 1 + 1 = 2! And by modern standards, the formal axiomatic system used there is inadequate; it's not formal enough! No wonder that Poincaré decried this as a kind of madness!

But I think that it was an interesting philosophical/intellectual exercise to refute this madness as convincingly as possible; somehow I ended up spending my life on that!

Of course, formalism fits the 20th century zeitgeist so well: Everything is meaningless, technical papers should never discuss ideas, only present the facts! A rule that I've done my best to ignore! As Vladimir Tasić says in his book *Mathematics and the Roots of Postmodern Thought,* a great deal of 20th-century philosophy seems enamored with formalism and then senses that Gödel has pulled the rug out from under it, and therefore truth is relative.—Or, as he puts it, it **could** have happened this way.—Tasić presents 20th century thought as, in effect, a dialogue between Hilbert and Poincaré . . . But truth is

[8]Mathematicians this century have nevertheless made the fatal mistake of gradually eliminating all words from their papers. That's another story, which I prefer not to discuss here, the Bourbakisation of 20th-century mathematics.

relative is **not** the correct conclusion. The correct conclusion is that Hilbert was wrong and that Poincaré was right: intuition cannot be eliminated from mathematics, or from human thought in general. And it's not all the same, all intuitions are not equally valid. Truth is **not** reinvented by each culture or each generation, it's not merely a question of fashion.

Let me repeat: formal axiomatic systems are a failure! Theorem proving algorithms **do not work.** One can publish papers about them, but they only prove trivial theorems. And in the case histories in this book, we've seen that the essence of math resides in its creativity, in imagining new concepts, in changing viewpoints, not in mindlessly and mechanically grinding away deducing all the possible consequences of a fixed set of rules and ideas.

Similarly, proving correctness of software using formal methods is hopeless. Debugging is done experimentally, by trial and error. And cautious managers insist on running a new system in parallel with the old one until they believe that the new system works.

Except—what we did in a marvelous project at IBM—for constantly eating your own cooking. We were constantly running the latest version of our software, constantly compiling our optimizing compiler through itself. By constantly using it ourselves, we got instantaneous feedback on the performance and on the design, which was constantly evolving. In my opinion that's the only way to develop a large piece of software: a totalitarian top-down approach cannot work. You have to design it as you go, not at the very beginning, before you write a single line of code.

And I feel that my experience in the real world debugging software yields a valuable philosophical lesson: Experimentation is the only way to "prove" that software is correct. Traditional mathematical proofs are only possible in toy worlds, not in the real world. The real world is too complicated. A physicist would say it like this: Pure math can only deal with the hydrogen atom. One proton, one electron, that's it! The quasi-empirical view of math may be controversial in the math community, but it is old hat in the software business. Programmers already have a quasi-empirical attitude to proof. Even though software is pure mind-stuff, not physical, programmers behave like physicists, not mathematicians, when it comes to debugging.

On the other hand, a way to minimize the debugging problem is to try at all costs to keep software intellectually manageable, as illustrated by my discussion of LISP. In our IBM project, I did this by rewriting all my code from scratch each time that I advanced in my understanding of the problem. I refused to use *ad hoc* code and tried instead to base everything as much as possible on clean, systematic mathematical algorithms. My goal was crystalline clarity. In my opinion, what counts most in a design is conceptual integrity, being faithful to an idea, not confusing the issue!

The previous paragraph is very pro-math. However, my design evolved, as I said, based on computer experiments, and the experimental method is used by physicists, not by mathematicians. It's too bad that mathematicians feel that way about experimentation!

Remember, I think that math is not that different from physics, for we must be willing to add new axioms:

Physics: laws → Computer → universe

Math: axioms → Computer → theorems

To get more out, put more in!

Another lesson I'd like you to take from this book is that everything is connected, all important ideas are—and that fundamental questions go back millennia and are **never** resolved. For example, the tension between the continuous and the discrete, or the tension between the world of ideas (math!) and the real world (physics, biology). You can find all this discussed in ancient Greece. And I suspect we could even trace it back to ancient Sumer, if more remained of Sumerian math than the scrap paper jottings on the clay tablets that are all we have, jottings that give hints of surprisingly sophisticated methods and of a love for calculation that seems to far outstrip any possible practical application.[9]

ON CREATIVITY

Having emphasized and re-emphasized the importance of creativity, it would be nice if I had a theory about it. Nope, but **I do have** some experience being creative. So let me try to share that with you.

The message of Gödel incompleteness, as I've said again and again in this book, is that a static fixed FAS cannot work. You have to add new information, new axioms, new concepts. Math is constantly evolving. The problem with current metamath is that it deals only with—it refutes—static FAS's. So where do new math ideas come from? Can we have a theory about that? A dynamic rather than static view of math, a dynamic rather than a static metamath, a kind of dynamic FAS perhaps?

Since I don't have that theory, I think an anecdotal approach might be best. This book is full of amazing case studies of new, unexpected math ideas that reduced the complicated to the obvious. And I've come up with a few of these ideas myself. How does it feel to do that?

Well, you can't find them if you don't look for them, if you don't **really believe** in them.

Is there some way to train for it, like a sport?! No, I don't think so! You have to be seized by a demon, and our society doesn't want too many people to be like that!

Let me describe what it feels like right now while I'm writing this book.

First of all, the ideas that I'm discussing seem very concrete, real and tangible to me. Sometimes they even feel more real than the peo-

[9]Did Sumer inherit its mathematics from an **even older** civilization—one more advanced than the ancient Greeks—that was destroyed by the glaciers, or when the glaciers suddenly melted, or by some other natural catastrophe? There is no way for such sophisticated computational techniques to appear out of nowhere, without antecedents.

ple around me. They certainly feel more real than newspapers, shopping malls and TV programs—those always give me a tremendous feeling of **unreality**! In fact, I only really feel alive when I'm working on a new idea, when I'm making love to a woman (which is also working on a new idea, the child we might conceive), or when I'm going up a mountain! It's intense, very intense.

When I'm working on a new idea I push everything else away. I stop swimming in the morning, I don't pay the bills, I cancel my doctor appointments. As I said, everything else becomes unreal! And I don't have to force myself to do this.

On the contrary, it's pure sensuality, pure pleasure. I put beautiful new ideas in the same category with beautiful women and beautiful art. To me it's like an amazing ethnic cuisine I've never tasted before.

I'm not depriving myself of anything, I'm not an ascetic. I don't look like an ascetic, do I?

And you can't force yourself to do it, any more than a man can force himself to make love to a woman he doesn't want.

The good moments are very, very good! Sometimes when I'm writing this I don't know where the ideas come from. I think that it can't be me, that I'm just a channel for ideas that want to be expressed.—But I **have** been concentrating on these questions for a long time.—I feel inspired, energized by the ideas. People may think that something's wrong with me, but I'm okay, I'm more than okay. It's pure enthusiasm! That's "God within" in Greek. Intellectual elation—like summiting on a high peak!

And I'm a great believer in the subconscious, in sleeping on it, in going to bed at 3 a.m. or 5 a.m. after working all night, and then getting up the next morning full of new ideas, ideas that come to you in waves while you're taking a bath, or having coffee. Or swimming laps. So mornings are very important to me, and I prefer to spend them at home. Routine typing and e-mail, I do in my office, not at home. And when I get too tired to stay in the office, then I print out the final version of the chapter I'm working on, bring it home—where there is **no** computer—and lie in bed for hours reading it, thinking about it, making corrections, adding stuff.

Sometimes the best time is lying in bed in the dark with my eyes closed, in a half dreamy, half awake state that seems to make it easier for new ideas, or new combinations of ideas, to emerge. I think of the subconscious as a chemical soup that's constantly making new combinations, and interesting combinations of ideas stick together, and eventually percolate up into full consciousness.—That's not too different from a biological population in which individuals fall in love and combine to produce new individuals.—My guess is that all this activity takes place at a molecular level—like DNA and information storage in the immune system—not at the cellular level. That's why the brain is so powerful, because that's where the real information processing is, at a molecular level. The cellular level, that's just the front end . . .

Yes, I believe in ideas, in the power of imagination and new ideas. And I don't believe in money or in majority views or the consensus. Even if all you are interested in is money, I think that new ideas are vital in the long run, which is why a commercial enterprise like IBM has a Research Division and has supported my work for so long. Thank you, IBM!

And I think that this also applies to human society.

I think that the current zeitgeist is very dangerous, because people are really desperate for their lives to be meaningful. They need to be creative, they need to help other people, they need to be part of a community, they need to be adventurous explorers, all things that tribal life provided. That's why much of the art in my home is from so-called "primitive" cultures like Africa.

So you're not going to be surprised to hear that I think that we desperately need new ideas about how human society should be organized, about what it's all for and how to live.

When I was a child, I read a science-fiction story by Chad Oliver. He was an anthropologist (from Greek *anthropos* = human being), not someone interested in hardware, or in spaceships, but in what was going on inside, in the human soul.

The story was called *Election Day,* and it was about a typical American family consisting of a working father, a stay-at-home mother, and two children, a boy and a girl.—That was the 1950s, remember!—Everything seemed normal, until you learned that the father, who was running for office, actually wasn't himself a candidate, no, he had created a social system. And the election wasn't electing people, it was voting on the choice of the next social system. And this wasn't really 1950s America, they had voted to live that way for a while, and then they were going to change and try a completely different system, right away, soon after election day!

What a remarkable idea!

And that's what I think we desperately need: Remarkable new ideas! Paradigm shifts! Meaning, intuition, creativity! We need to reinvent ourselves!

Thank you for reading this book and for taking this journey with me!

READING A NOTE IN THE JOURNAL *NATURE* I LEARN

by Robert M. Chute

Omega, Omega-like, and Computably Enumerable
Random Real Numbers all may be
a single class.

Should I be concerned? Is this the sign
of a fatal fault line
in the logic of our world?

You shouldn't worry, I'm told,
about such things, but how to be indifferent
if you don't understand?

Standing here, we do not sense
any tectonic rearrangement
beneath our feet.

Despite such reassurance I feel
increasing dis-ease. Randomness is rising
around my knees.

Remember how we felt when we learned
that one infinity
could contain another?

Somewhere between the unreachable
and the invisible
I had hoped for an answer.[10]

[10]From *Beloit Poetry Journal,* Spring 2000 (Vol. 50, No. 3, p. 8). The note in *Nature* in question is C. S. Calude, G. J. Chaitin, "Randomness Everywhere," *Nature,* 22 July 1999 (Vol. 400, pp. 319–320).

MATH POEM

by Marion D. Cohen[11]

Someone wrote a book called *The Joy of Math.*
Maybe I'll write a book called *The Pathos of Math.*
For through the night I wander
between intuition and calculation
between examples and counter-examples
between the problem itself and what it has led to.
I find special cases with no determining vertices.
I find special cases with only determining vertices.
I weave in and out.
I rock to and fro.
I am the wanderer
with a lemma in every port.

[11]From her collection of math poetry *Crossing the Equal Sign* at http://mathwoman .com. Originally published in the April 1999 *American Mathematical Monthly.*

FURTHER READING—BOOKS/PLAYS/MUSICALS ON RELATED TOPICS

Instead of providing an infinitely long bibliography, I decided to concentrate mostly on **recent** books that caught my eye.

- Stephen Wolfram, *A New Kind of Science,* Wolfram Media, 2002. [A book concerned with many of the topics discussed here, but with quite a different point of view.]
- Joshua Rosenbloom and Joanne Sydney Lesser, *Fermat's Last Tango. A New Musical,* York Theatre Company, New York City, 2000. CD: Original Cast Records OC-6010, 2001. DVD: Clay Mathematics Institute, 2001. [A humorous and playful presentation of one mathematician's obsession with math.]
- David Foster Wallace, *Everything and More: A Compact History of Infinity,* Norton, 2003. [An American writer's take on mathematics.]
- Dietmar Dath, *Höhenrausch. Die Mathematik des XX. Jahrhunderts in zwanzig Gehirnen,* Eichborn, 2003. [A German writer's reaction to modern math and modern mathematicians.]
- John L. Casti, *The One True Platonic Heaven. A Scientific Fiction on the Limits of Knowledge,* Joseph Henry Press, 2003. [Gödel, Einstein and von Neumann at the Princeton Institute for Advanced Study.]
- Apostolos Doxiadis, *Incompleteness, A Play and a Theorem,* http://www.apostolosdoxiadis.com. [A play about Gödel from the author of *Uncle Petros and Goldbach's Conjecture.*]
- Mary Terall, *The Man Who Flattened the Earth. Maupertuis and the Sciences in the Enlightenment,* University of Chicago Press,

2002. [Newton vs. Leibniz, the generation after; a portrait of an era.]

- Isabelle Stengers, *La Guerre des sciences aura-t-elle lieu? Scientifiction,* Les Empêcheurs de penser en rond/Le Seuil, 2001. [A play about Newton vs. Leibniz.]
- Carl Djerassi, David Pinner, *Newton's Darkness. Two Dramatic Views,* Imperial College Press, 2003. [Two plays, one about Newton vs. Leibniz.]
- Neal Stephenson, *Quicksilver,* Morrow, 2003. [A science fiction novel about Newton vs. Leibniz. Volume 1 of 3.]
- David Ruelle, *Chance and Chaos,* Princeton Science Library, Princeton University Press, 1993. [A physicist's take on randomness.]
- Cristian S. Calude, *Information and Randomness,* Springer-Verlag, 2002. [A mathematician's take on randomness.]
- James D. Watson, Andrew Berry, *DNA: The Secret of Life,* Knopf, 2003. [The role of information in biology.]
- Tom Siegfried, *The Bit and the Pendulum. From Quantum Computing to M Theory—The New Physics of Information,* Wiley, 2000. [The role of information in physics.]
- Tor Nørretranders, *The User Illusion. Cutting Consciousness Down to Size,* Viking, 1998. [Information theory and the mind.]
- Hans Christian von Baeyer, *Information: The New Language of Science,* Weidenfeld & Nicholson, 2003.
- Marcus du Sautoy, *The Music of the Primes: Searching to Solve the Greatest Mystery in Mathematics,* HarperCollins, 2003. [Of the recent crop of books on the Reimann hypothesis, the only one that considers the possibility that Gödel incompleteness might apply.]
- Douglas S. Robertson, *The New Renaissance: Computers and the Next Level of Civilization,* Oxford University Press, 1998.
- Douglas S. Robertson, *Phase Change: The Computer Revolution in Science and Mathematics,* Oxford University Press, 2003.
  [These two books by Robertson discuss the revolutionary social transformations provoked by improvements in information transmission technology.]
- John Maynard Smith, Eörs Szathmáry, *The Major Transitions in Evolution,* Oxford University Press, 1998.

- John Maynard Smith, Eörs Szathmáry, *The Origins of Life: From the Birth of Life to the Origin of Language,* Oxford University Press, 1999.

  [These two books by Maynard Smith and Szathmáry discuss evolutionary progress in terms of radical improvements in the representation of biological information.]
- John D. Barrow, Paul C. W. Davies, Charles L. Harper, Jr., *Science and Ultimate Reality: Quantum Theory, Cosmology, and Complexity,* Cambridge University Press, 2004. [Essays in honor of John Wheeler.]
- Gregory J. Chaitin, *Conversations with a Mathematician,* Springer-Verlag, 2002. [Previous book by the author of this book.]
- Robert Wright, *Three Scientists and Their Gods: Looking for Meaning in an Age of Information,* HarperCollins, 1989. [On Fredkin's thesis that the universe is a computer.]
- Jonathan M. Borwein, David H. Bailey, *Mathematics by Experiment: Plausible Reasoning in the 21st Century,* A. K. Peters, 2004. [How to discover new mathematics. Volume 1 of 2.]
- Vladimir Tasić, *Mathematics and the Roots of Postmodern Thought,* Oxford University Press, 2001. [Where I learnt about Borel's know-it-all number.]
- Thomas Tymoczko, *New Directions in the Philosophy of Mathematics,* Princeton University Press, 1998.
- Newton C. A. da Costa, Steven French, *Science and Partial Truth,* Oxford University Press, 2003.
- Eric B. Baum, *What Is Thought?,* MIT Press, 2004.

Appendix I

COMPUTERS, PARADOXES AND THE FOUNDATIONS OF MATHEMATICS

Some great thinkers of the 20th century
have shown that even in the austere world of mathematics,
incompleteness and randomness are rife

from *American Scientist,* March–April 2002, pp. 164–171

Everyone knows that the computer is a very practical thing. In fact, computers have become indispensable to running a modern society. But what even computer experts don't remember is that—I exaggerate only slightly—the computer was invented in order to help to clarify a philosophical question about the foundations of mathematics. Surprising? Yes indeed.

This amazing story begins with David Hilbert, a well-known German mathematician who at the start of the 20th century proposed formalizing completely all of mathematical reasoning. It turned out that you can't formalize mathematical reasoning, so in one sense his idea was a tremendous failure. Yet in another way, Hilbert's idea was a success, because formalism has been one of the biggest boons of the 20th century—not for mathematical reasoning or deduction, but for programming, for calculating, for computing. This is a forgotten piece of intellectual history.

I will relate that history here without delving into mathematical details. So it will be impossible to fully explain the work of the relevant contributors, who include Bertrand Russell, Kurt Gödel and Alan Turing. Still, a patient reader should be able to glean the essence of their arguments and see what inspired some of my own ideas about the randomness inherent in mathematics.

RUSSELL'S LOGICAL PARADOXES

Let me start with Bertrand Russell, a mathematician who later turned into a philosopher and finally into a humanist. Russell is key because he discovered some disturbing paradoxes in logic itself. That is, he found cases where reasoning that seems to be sound leads to contradictions. Russell was tremendously influential in spreading the idea that these contradictions constituted a serious crisis and had to be resolved somehow.

The paradoxes that Russell discovered attracted a great deal of attention in mathematical circles, but strangely enough only one of them ended up with his name on it. To understand the Russell paradox, consider the set of all sets that are not members of themselves. Then ask, "Is this set a member of itself?" If it is a member of itself, then it shouldn't be, and vice versa.

The set of all sets in the Russell paradox is like the barber in a small, remote town who shaves all the men who don't shave themselves. That description seems pretty reasonable, until you ask, "Does the barber shave himself?" He shaves himself if and only if he doesn't shave himself. Now you may say, "Who cares about this hypothetical barber? This is just silly wordplay!" But when you're dealing with the mathematical concept of a set, it's not so easy to dismiss a logical problem.

The Russell paradox is a set-theoretic echo of an earlier paradox, one that was known to the ancient Greeks. It is often called the Epimenides paradox or the paradox of the liar. The essence of the problem is this: Epimenides was said to exclaim, "This statement is false!" Is it false? If his statement is false, that means that it must be true. But if it's true, it's false. So whatever you assume about his veracity, you're in trouble. A two-sentence version of the paradox goes like this: "The following statement is true. The preceding statement is false." Either statement alone is okay, but combined they make no sense. You might dismiss such conundrums as meaningless word games, but some of the great minds of the 20th century took them very seriously.

One of the reactions to the crisis of logic was Hilbert's attempt to escape into formalism. If one gets into trouble with reasoning that seems okay, the solution is to use symbolic logic to create an artificial language and be very careful to specify the rules so that contradictions don't crop up. After all, everyday language is ambiguous—you never know what a pronoun refers to.

HILBERT'S RESCUE PLAN

Hilbert's idea was to create a perfect artificial language for reasoning, for doing mathematics, for deduction. Hence, he stressed the importance of the axiomatic method, whereby one works from a set of basic postulates (axioms) and well-defined rules of deduction to derive valid theorems. The notion of doing mathematics that way goes back to the ancient Greeks and particularly to Euclid and his geometry, which is a beautifully clear mathematical system.

In other words, Hilbert's intention was to be completely precise about the rules of the game—about the definitions, the elementary concepts, the grammar and the language—so that everyone could agree on how mathematics should be done. In practice it would be too much work to use such a formal axiomatic system for developing new mathematics, but it would be philosophically significant.

Hilbert's proposal seemed fairly straightforward. After all, he was just following the formal traditions in mathematics, drawing from a long history of work by Leibniz, Boole, Frege and Peano. But he wanted to go all the way to the very end and formalize *all* of mathematics. The big surprise is that it turned out that this could not be done. Hilbert was wrong—but wrong in a tremendously fruitful way, because he had asked a very good question. In fact, by asking this question he created an entirely new discipline called *meta-mathematics,* an introspective field of math in which you study what mathematics can or cannot achieve.

The basic concept is this: Once you entomb mathematics in an artificial language *à la* Hilbert, once you set up a completely formal axiomatic system, then you can forget that it has any meaning and just look at it as a game played with marks on paper that enables you to deduce theorems from axioms. Of course, the reason one does mathematics is because it has meaning. But if you want to be able to study mathematics using mathematical methods, you have to crystallize out the meaning and just examine an artificial language with completely precise rules.

What kind of questions might you ask? Well, one question is whether one can prove, say, that $0 = 1$. (We can hope not.) Indeed, for any statement, call it *A,* you can ask if it's possible to prove either *A* or the opposite of *A*. A formal axiomatic system is considered complete if you either prove that *A* is true, or prove that it's false.

Hilbert envisioned creating rules so precise that any proof could always be submitted to an unbiased referee, a mechanical procedure that would say, "This proof obeys the rules," or perhaps, "On line 4 there's a misspelling" or, "This thing on line 4 that supposedly follows from line 3, actually doesn't." And that would be the end; no appeal.

His idea was not that mathematics should actually be done this way, but rather that if you could take mathematics and do it this way, you could then use mathematics to study the power of mathematics. And Hilbert thought that he was actually going to be able to accomplish that feat. So you can imagine just how very, very shocking it was in 1931 when an Austrian mathematician named Kurt Gödel showed that Hilbert's rescue plan wasn't at all reasonable. It could never be carried out, even in principle.

GÖDEL'S INCOMPLETENESS

Gödel exploded Hilbert's vision in 1931 while on the faculty of the University of Vienna, although he originally came from what is now called the Czech Republic, from the city of Brno. (It was then part of the Austro-Hungarian empire.) Later Gödel was to join Einstein at the Institute for Advanced Study in Princeton.

Gödel's amazing discovery is that Hilbert was dead wrong: There is, in fact, no way to have a formal axiomatic system for all of mathematics in which it is crystal clear whether something is correct or not. More precisely, what Gödel discovered was that the plan fails even if you just try to deal with elementary arithmetic, with the numbers 0, 1, 2, 3, ... and with addition and multiplication.

Any formal system that tries to contain the whole truth and nothing but the truth about addition, multiplication and the numbers 0, 1, 2, 3, ... will have to be incomplete. Actually, it will either be inconsistent or incomplete. So if you assume that it only tells the truth, then it won't tell the whole truth. In particular, if you assume that the axioms and rules of deduction don't allow you to prove false theorems, then there will be true theorems that you cannot prove.

Gödel's incompleteness proof is very clever. It's very paradoxical. It almost looks crazy. Gödel starts in effect with the paradox of the liar: the statement, "I'm false!," which is neither true nor false. Actually what Gödel does is to construct a statement that says of itself, "I'm unprovable!" Now if you can construct such a statement in elementary number theory, in arithmetic, a mathematical statement that describes itself, you've got to be very clever—but if you *can* do it, it's easy to see that you're in trouble. Why? Because if that statement is provable, it is by necessity false, and you're proving false results. If it's unprovable, as it says of itself, then it's true, and mathematics is incomplete.

Gödel's proof involves many complicated technical details. But if you look at his original paper, you find something that looks a lot like LISP programming in it. That is because Gödel's proof involves defining a great many functions recursively, functions dealing with lists—precisely what LISP is all about. So even though there were no computers or programming languages in 1931, with the benefit of hindsight you can clearly see a programming language at the core of Gödel's original paper.

Another famous mathematician of that era, John von Neumann (who, incidentally, was instrumental in encouraging the creation of computer technology in the United States), appreciated Gödel's insight immediately. It had never occurred to von Neumann that Hilbert's plan was unsound. So Gödel was not only tremendously clever, he had the courage to imagine that Hilbert might be wrong.

Many people saw Gödel's conclusion as absolutely devastating: All of traditional mathematical philosophy ended up in a heap on the floor. In 1931, however, there were also a few other problems to worry about in Europe. There was a major depression, and a war was brewing.

TURING'S MACHINE

The next major step forward came five years later, in England, when Alan Turing discovered uncomputability. Recall that Hilbert had said that there should be a "mechanical procedure" to decide if a proof obeys the rules or not. Hilbert never clarified what he meant by a mechanical procedure. Turing essentially said, "What you really mean is a machine" (a machine of a kind that we now call a Turing machine).

Turing's original paper contains a programming language, just as Gödel's paper does, or what we would now call a programming language. But these two programming languages are very different. Turing's isn't a high-level language like LISP; it's more like a machine language, the raw code of ones and zeros that are fed to a computer's central processor. Turing's invention of 1936 is, in fact, a horrible machine language, one that nobody would want to use today, because it's too rudimentary.

Although Turing's hypothetical computing machines are very simple and their machine language rather primitive, they're very flexible. In his 1936 paper, Turing claims that such a machine should be able to perform any computation that a human being can carry out.

Turing's train of thought now takes a very dramatic turn. What, he asks, is *impossible* for such a machine? What can't it do? And he immediately finds a problem that no Turing machine can solve: the halting problem. This is the problem of deciding in advance whether a Turing machine (or a computer program) will eventually find its desired solution and halt.

If you allow a time limit, it's very easy to solve this problem. Say that you want to know whether a program will halt within a year. Then you just run it for a year, and it either halts or doesn't. What Turing showed is that you get in terrible trouble if you impose no time limit, if you try to deduce whether a program will halt without just running it.

Let me outline Turing's reasoning: Suppose that you *could* write a computer program that checks whether any given computer program eventually halts. Call it a termination tester. In theory, you feed it a program, and it would spew out an answer: "Yes, this program will terminate," or, "No, it will go off spinning its wheels in some infinite loop and never come to a halt." Now create a second program that uses the termination tester to evaluate some program. If the program under investigation terminates, have your new program arranged so that it goes into an infinite loop. Here comes the subtle part: Feed your new program a copy of itself. What does it do?

Remember, you've written this new program so that it will go into an infinite loop if the program under test terminates. But here it is *itself* the program under test. So if it terminates, it goes off in an infinite loop, which means it doesn't terminate—a contradiction. Assuming the opposite outcome doesn't help: If it doesn't terminate, the termination tester will indicate this, and the program will not go into an infinite loop, thus terminating. The paradox led Turing to conclude that a general purpose termination tester couldn't be devised.

The interesting thing is that Turing immediately deduced a corollary: If there's no way to determine in advance by a calculation whether a program will halt, there also cannot be any way to decide it in advance using reasoning. No formal axiomatic system can enable you to deduce whether a program will eventually halt. Why? Because if you could use a formal axiomatic system in this way, that would give you the means to calculate in advance whether a program will halt or not. And that's impossible, because you get into a paradox like, "This statement is false!" You can create a program that halts if and only if it doesn't halt. The paradox is similar to what Gödel found in his investigation of number theory. (Recall he was looking at nothing more complicated than 0, 1, 2, 3, ... and addition and multiplication.) Turing's coup is that he showed that *no* formal axiomatic system can be complete.

After World War II broke out, Turing began working on cryptography, von Neumann started working on how to calculate atom-bomb detonations, and people forgot about the incompleteness of formal axiomatic systems for a while.

RANDOMNESS IN MATHEMATICS

The generation of mathematicians who were concerned with these deep philosophical questions basically disappeared with World War II. Then I showed up on the scene.

In the late 1950s, when I was a youngster, I read an article on Gödel and incompleteness in *Scientific American.* Gödel's result fascinated me, but I couldn't really understand it; I thought there was something fishy. As for Turing's approach, I appreciated that it went much deeper, but I still wasn't satisfied. This is when I got a funny idea about randomness.

When I was a kid, I also read a lot about another famous intellectual issue, not the foundations of mathematics but the foundations of physics—about relativity theory and cosmology and even more often about quantum mechanics. I learned that when things are very small the physical world behaves in a completely crazy way. In fact, things are random—intrinsically unpredictable. I was reading about all of this, and I started to wonder whether there was also randomness in pure mathematics. I began to suspect that maybe this was the real reason for incompleteness.

A case in point is elementary number theory, where there are some very difficult questions. Consider the prime numbers. Individual prime numbers behave in a very unpredictable way, if you're interested in their detailed structure. It's true that there are statistical patterns. There's a thing called the prime number theorem that predicts fairly accurately the overall average distribution of the primes. But as for the detailed distribution of individual prime numbers, that looks pretty random.

So I began to think that maybe the inherent randomness in mathematics provided a deeper reason for all this incompleteness. In the mid-1960s I, and independently A. N. Kolmogorov in the U.S.S.R., came up with some new ideas, which I like to call algorithmic information theory. That name makes it sound very impressive, but the basic idea is simple: It's just a way to measure computational complexity.

One of the first places that I heard about the idea of computational complexity was from von Neumann. Turing considered the computer as a mathematical concept—a perfect computer, one that never makes mistakes, one that has as much time and space as it needs to do its work. After Turing came up with this idea, the next logical step for a mathematician was to study the time needed to do a calculation—a measure of its complexity. Around 1950 von Neumann highlighted the importance of the time complexity of computations, and that is now a well-developed field.

My idea was not to look at the time, even though from a practical point of view time is very important. My idea was to look at the *size* of computer programs, at the amount of information that you have to give a computer to get it to perform a given task. Why is that interesting? Because program-size complexity connects with the notion of entropy in physics.

Recall that entropy played a particularly crucial role in the work of the famous 19th-century physicist Ludwig Boltzmann, and it comes up in the fields of statistical mechanics and thermodynamics. Entropy measures the degree of disorder, chaos, randomness, in a physical system. A crystal has low entropy, and a gas (say, at room temperature) has high entropy.

Entropy is connected with a fundamental philosophical question: Why does time run in just one direction? In everyday life, there is, of course, a great difference between going backward and forward in time. Glasses break, but they don't reassemble spontaneously. Similarly, in Boltzmann's theory, entropy has to increase—the system has to get more and more disordered. This is the well-known Second Law of Thermodynamics.

Boltzmann's contemporaries couldn't see how to deduce that result from Newtonian physics. After all, in a gas, where the atoms bounce around like billiard balls, each interaction is reversible. If you were somehow able to film a small portion of a gas for a brief time, you couldn't tell whether you were seeing the movie run forward or backward. But Boltzmann's gas theory says that there *is* an arrow of time—a system will start off in an ordered state and will end up in a very mixed up disordered state. There's even a scary expression for the final condition, "heat death."

The connection between my ideas and Boltzmann's theory comes about because the size of a computer program is analogous to the degree of disorder of a physical system. A gas might take a large program to say where all its atoms are located, whereas a crystal doesn't require as big a program at all, because of its regular structure. Entropy and program-size complexity are thus closely related.

This concept of program-size complexity is also connected with the philosophy of the scientific method. Ray Solomonoff (a computer scientist then working for the Zator Company in Cambridge, Massachusetts) proposed this idea at a conference in 1960, although I only learned of his work after I came up with some very similar ideas on my own a few years later. Just think of Occam's razor, the idea that the simplest theory is best. Well, what's a theory? It's a computer program for predicting observations. And the statement that the simplest theory is best translates into saying that a concise computer program constitutes the best theory.

What if there is no concise theory? What if the most concise program for reproducing a given body of experimental data is the same size as the data set? Then the theory is no good—it's cooked up—and the data are incomprehensible, random. A theory is good only to the extent that it compresses the data into a much smaller set of theoretical assumptions and rules for deduction.

So you can define randomness as something that cannot be compressed at all. The only way to describe a completely random object or number to someone else is to present it and say, "This is it." Because it has no structure or pattern, there is no more concise description. At the other extreme is an object or number that has a very regular pattern. Perhaps you can describe it by saying that it is a million repetitions of 01, for example. This is a very big object with a very short description.

My idea was to use program-size complexity to define randomness. And when you start looking at the size of computer programs—when you begin to think about this notion of program-size or information complexity instead of run-time complexity—then an interesting thing happens: Everywhere you turn, you find incompleteness. Why? Because the very first question you ask in my theory gets you into trouble. You measure the complexity of something by the size of the smallest computer program for calculating it. But how can you be sure that what you have is the smallest computer program possible? The answer is that you can't. This task escapes the power of mathematical reasoning, amazingly enough.

Showing why this is so gets somewhat involved, so I'm just going to quote the actual result, which is one of my favorite statements of incompleteness: If you have *n* bits of axioms, you can never prove that a program is the smallest possible if it is more than *n* bits long. That is, you get into trouble with a program if it's larger than a computerized version of the axioms—or more precisely, if it's larger than the size of the proof-checking program for the axioms and the associated rules of deduction.

So it turns out that you cannot in general calculate program-size complexity, because to determine the program-size complexity of something is to know the size of the most concise program that calculates it. You can't do that if the program is larger than the axioms. If there are *n* bits of axioms, you can never determine the program-size complexity of anything that has more than *n* bits of complexity, which means almost everything.

Let me explain why I claim that. The sets of axioms that mathematicians normally use are fairly concise, otherwise no one would believe in them. In practice, there's this vast world of mathematical truth out there—an infinite amount of information—but any given set of axioms only captures a tiny, finite amount of this information. That, in a nutshell, is why Gödel incompleteness is natural and inevitable rather than mysterious and complicated.

WHERE DO WE GO FROM HERE?

This conclusion is very dramatic. In only three steps one goes from Gödel, where it seems shocking that there are limits to reasoning, to Turing, where it looks much more reasonable, and then to a consideration of program-size complexity, where incompleteness, the limits of mathematics, just hits you in the face.

People often say to me, "Well, this is all very well and good. Algorithmic information theory is a nice theory, but give me an example of a specific result that you think escapes the power of mathematical reasoning." For years, one of my favorite answers was, "Perhaps Fermat's last theorem." But a funny thing happened: In 1993 Andrew Wiles came along with a proof. There was a misstep, but now everyone is convinced that the proof is correct. So there's the problem. Algorithmic information theory shows that there are lots of things that you can't prove, but it cannot reach a conclusion for individual mathematical questions.

How come, in spite of incompleteness, mathematicians are making so much progress? These incompleteness results certainly have a pessimistic feeling about them. If you take them at face value, it would seem that there's no way to progress, that mathematics is impossible. Fortunately for those of us who do mathematics, that doesn't seem to be the case. Perhaps some young metamathematician of the next generation will prove why this has to be so.

BIBLIOGRAPHY

- Casti, J. L., and W. DePauli. 2000. *Gödel: A Life of Logic.* Cambridge, Mass.: Perseus Publishing.
- Chaitin, G. J. 1975. Randomness and mathematical proof. *Scientific American* 232(5):47–52.
- Chaitin, G. J. 1988. Randomness in arithmetic. *Scientific American* 259(1):80–85.
- Hofstadter, D. R. 1979. *Gödel, Escher, Bach: an Eternal Golden Braid.* New York: Basic Books.
- Nagel, E., and J. R. Newman. 1956. Gödel's proof. *Scientific American* 194(6):71–86.
- Nagel, E., and J. R. Newman. 1958. *Gödel's Proof.* New York: New York University Press.

Appendix II

ON THE INTELLIGIBILITY OF THE UNIVERSE AND THE NOTIONS OF SIMPLICITY, COMPLEXITY AND IRREDUCIBILITY

from *Grenzen und Grenzüberschreitungen, XIX. Deutscher Kongress für Philosophie, Bonn, 23.-27. September 2002, Vorträge und Kolloquien,* Herausgegeben von Wolfram Hogrebe in Verbindung mit Joachim Bromand, Akademie Verlag, Berlin, 2004, pp. 517–534

Abstract: We discuss views about whether the universe can be rationally comprehended, starting with Plato, then Leibniz, and then the views of some distinguished scientists of the previous century. Based on this, we defend the thesis that comprehension is compression, i.e., explaining many facts using few theoretical assumptions, and that a theory may be viewed as a computer program for calculating observations. This provides motivation for defining the complexity of something to be the size of the simplest theory for it, in other words, the size of the smallest program for calculating it. This is the central idea of algorithmic information theory (AIT), a field of theoretical computer science. Using the mathematical concept of program-size complexity, we exhibit irreducible mathematical facts, mathematical facts that cannot be demonstrated using any mathematical theory simpler than they are. It follows that the world of mathematical ideas has infinite complexity and is therefore not fully comprehensible, at least not in a static fashion. Whether the physical world has finite or infinite complexity remains to be seen. Current science believes that the world contains randomness, and is therefore also infinitely complex, but a deterministic universe that simulates randomness via pseudo-randomness is also a possibility, at least according to recent highly speculative work of S. Wolfram.

> "Nature uses only the longest threads to weave her patterns, so that each small piece of her fabric reveals the organization of the entire tapestry."
>
> —Feynman, *The Character of Physical Law,* 1965, at the very end of Chapter 1, "The Law of Gravitation." [An updated version of this chapter would no doubt include a discussion of the infamous astronomical missing mass problem.]

> "The most incomprehensible thing about the universe is that it is comprehensible."
>
> —Attributed to Einstein. The original source, where the wording is somewhat different, is Einstein, "Physics and Reality," 1936, reprinted in Einstein, *Ideas and Opinions,* 1954.

It's a great pleasure for me to speak at this meeting of the German Philosophical Society. Perhaps it's not generally known that at the end of his life my predecessor Kurt Gödel was obsessed with Leibniz. [See Menger, *Reminiscences of the Vienna Circle and the Mathematical Colloquium,* 1994.] Writing this paper was for me a voyage of discovery—of the depth of Leibniz's thought! Leibniz's power as a philosopher is informed by his genius as a mathematician; as I'll explain, some of the key ideas of AIT are clearly visible in embryonic form in his 1686 *Discourse on Metaphysics.*

[Einstein actually wrote

"Das ewig Unbegreifliche an der Welt ist ihre Begreiflichkeit."

Translated word for word, this is

"The eternally incomprehensible about the world is its comprehensibility."

But I prefer the version given above, which emphasizes the paradox.]

I. PLATO'S *TIMAEUS*—THE UNIVERSE IS INTELLIGIBLE. ORIGINS OF THE NOTION OF SIMPLICITY: SIMPLICITY AS SYMMETRY [BRISSON, MEYERSTEIN 1991]

"[T]his is the central idea developed in the *Timaeus:* the order established by the demiurge in the universe becomes manifest as the symmetry found at its most fundamental level, a symmetry which makes possible a mathematical description of such a universe."

—Brisson, Meyerstein, *Inventing the Universe,* 1995 (1991 in French). This book discusses the cosmology of Plato's *Timaeus,* modern cosmology and AIT; one of their key insights is to identify symmetry with simplicity.

According to Plato, the world is rationally understandable because it has structure. And the universe has structure, because it is a work of art created by a God who is a mathematician. Or, more abstractly, the structure of the world consists of God's thoughts, which are mathematical. The fabric of reality is built out of eternal mathematical truth [Brisson, Meyerstein, *Inventer l'Univers,* 1991].

Timaeus postulates that simple, symmetrical geometrical forms are the building blocks for the universe: the circle and the regular solids (cube, tetrahedron, icosahedron, dodecahedron, octahedron).

What was the evidence that convinced the ancient Greeks that the world is comprehensible? Partly it was the beauty of mathematics, particularly geometry and number theory, and partly the Pythagorean work on the physics of stringed instruments and musical tones, and in astronomy, the regularities in the motions of the planets and the starry heavens and eclipses. Strangely enough, mineral crystals, whose symmetries magnify enormously quantum-mechanical symmetries that are found at the atomic and molecular level, are never mentioned.

What is our current cosmology?

Since the chaos of everyday existence provides little evidence of simplicity, biology is based on chemistry is based on physics is based on high-energy or particle physics. The attempt to find underlying simplicity and pattern leads reductionist modern science to break things into smaller and smaller components in an effort to find the underlying simple building blocks.

And the modern version of the cosmology of *Timaeus* is the application of symmetries or group theory to understand sub-atomic particles (formerly called elementary particles), for example, Gell-Mann's eightfold way, which predicted new particles. This work classifying the "particle zoo" also resembles Mendeleev's periodic table of the elements that organizes their chemical properties so well. (For more on this, see the essay by Freeman Dyson on "Mathematics in the Physical Sciences" in COSRIMS, *The Mathematical Sciences,* 1969. This is an article of his that was originally published in *Scientific American.*)

And modern physicists have also come up with a possible answer to the Einstein quotation at the beginning of this paper. Why do they think that the universe is comprehensible? They invoke the so-called "anthropic principle" [Barrow, Tipler, *The Anthropic Cosmological Principle,* 1986], and declare that we would not be here to ask this question unless the universe had enough order for complicated creatures like us to evolve!

Now let's proceed to the next major step in the evolution of ideas on simplicity and complexity, which is a stronger version of the Platonic creed due to Leibniz.

II. WHAT DOES IT MEAN FOR THE UNIVERSE TO BE INTELLIGIBLE? LEIBNIZ'S DISCUSSION OF SIMPLICITY, COMPLEXITY AND LAWLESSNESS [WEYL 1932]

"As for the simplicity of the ways of God, this holds properly with respect to his means, as opposed to the variety, richness, and abundance, which holds with respect to his ends or effects.

"But, when a rule is extremely complex, what is in conformity with it passes for irregular. Thus, one can say, in whatever manner God might have created the world, it would always have been regular and in accordance with a certain general order. But God has chosen the most perfect world, that is, the one which is at the same time the simplest in hypotheses and the richest in phenomena, as might be a line in geometry whose construction is easy and whose properties and effects are extremely remarkable and widespread."

—Leibniz, *Discourse on Metaphysics,* 1686, Sections 5–6, from Leibniz, *Philosophical Essays,* edited and translated by Ariew and Garber, 1989, pp. 38–39.

"The assertion that nature is governed by strict laws is devoid of all content if we do not add the statement that it is governed by mathematically simple laws . . . That the notion of law becomes empty when an arbitrary complication is permitted was already pointed out by Leibniz in his *Metaphysical Treatise* [*Discourse on Metaphysics*]. Thus simplicity becomes a working principle in the natural sciences . . . The astonishing thing is not that there exist natural laws, but that the further the analysis proceeds, the finer the details, the finer the elements to which the phenomena are reduced, the simpler—and not the more complicated, as one would originally expect—the fundamental relations become and the more exactly do they describe the actual occurrences. But this circumstance is apt to weaken the metaphysical power of determinism, since it makes the meaning of natural law depend on the fluctuating distinction between mathematically simple and complicated functions or classes of functions."

—Hermann Weyl, *The Open World, Three Lectures on the Metaphysical Implications of Science,* 1932, pp. 40–42. See a similar discussion on pp. 190–191 of Weyl, *Philosophy of Mathematics and Natural Science,* 1949, Section 23A, "Causality and Law." [This is a remarkable anticipation of my definition of "algorithmic randomness," as a set of observations that only has what Weyl considers to be unacceptable theories, ones that are as complicated as the observations themselves, without any "compression."]

> "Weyl said, not long ago, that 'the problem of simplicity is of central importance for the epistemology of the natural sciences.' Yet it seems that interest in the problem has lately declined; perhaps because, especially after Weyl's penetrating analysis, there seemed to be so little chance of solving it."
>
> —Weyl, *Philosophy of Mathematics and Natural Science,* 1949, p. 155, quoted in Popper, *The Logic of Scientific Discovery,* 1959, Chapter VII, "Simplicity," p. 136.

In his novel *Candide,* Voltaire ridiculed Leibniz, caricaturing Leibniz's subtle views with the memorable phrase **"this is the best of all possible worlds."** Voltaire also ridiculed the efforts of Maupertius to develop a physics in line with Leibniz's views, one based on a principle of least effort.

Nevertheless versions of least effort play a fundamental role in modern science, starting with Fermat's deduction of the laws for reflection and refraction of light from a principle of least time. This continues with the Lagrangian formulation of mechanics, stating that the actual motion minimizes the integral of the difference between the potential and the kinetic energy. And least effort is even important at the current frontiers, such as in Feynman's path integral formulation of quantum mechanics (electron waves) and quantum electro-dynamics (photons, electromagnetic field quanta). (See the short discussion of minimum principles in Feynman, *The Character of Physical Law,* 1965, Chapter 2, "The Relation of Mathematics to Physics." For more information, see *The Feynman Lectures on Physics,* 1963, Vol. 1, Chapter 26, "Optics: The Principle of Least Time," Vol. 2, Chapter 19, "The Principle of Least Action.")

However, all this modern physics refers to versions of least effort, not to ideas, not to information, and not to complexity—which are more closely connected with Plato's original emphasis on symmetry and intellectual simplicity = intelligibility. An analogous situation occurs in theoretical computer science, where work on computational complexity is usually focused on time, not on the complexity of ideas or information. Work on time complexity is of great practical value, but I believe that the complexity of ideas is of greater conceptual significance. Yet another example of the effort/information divide is the fact that I am interested in the irreducibility of ideas (see Sections V and VI), while Stephen Wolfram (who is discussed later in this section) instead emphasizes time irreducibility, physical systems for which there are no predictive short-cuts and the fastest way to see what they do is just to run them.

Leibniz's doctrine concerns more than "least effort," it also implies that the ideas that produce or govern this world are as beautiful and as simple as possible. In more modern terms, God employed the smallest possible amount of intellectual material to build the world, and the laws of physics are as simple and as beautiful as they can be and allow us, intelligent beings, to evolve. [This is a kind of "anthropic principle," the attempt to deduce things about the universe from the fact that we are here and able to look at it.] The belief in this Leibnizean doctrine lies behind the continuing reductionist efforts of high-energy physics (particle physics) to find the ultimate components of reality. The continuing vitality of this Leibnizean doctrine also lies behind astrophysicist John Barrow's emphasis in his "Theories of Everything" essay on finding the minimal TOE that explains the universe, a TOE that is as simple as possible, with no redundant elements (see Section VII below).

Important point: To say that the fundamental laws of physics must be simple does not at all imply that it is easy or fast to deduce from them how the world works, that it is quick to make predictions from the basic laws. The *apparent complexity* of the world we live in—a phrase that is constantly repeated in Wolfram, *A New Kind of Science,* 2002—then comes from the long deductive path from the basic laws to the level of our experience. [It could also come from the complexity of the initial conditions, or from coin-tossing, i.e., randomness.] So again, I claim that minimum information is more important than minimum time, which is why in Section IV I do not care how long a minimum-size program takes to produce its output, nor how much time it takes to calculate experimental data using a scientific theory.

More on Wolfram: In *A New Kind of Science,* Wolfram reports on his systematic computer search for simple rules with very complicated consequences, very much in the spirit of Leibniz's remarks above. First Wolfram amends the Pythagorean insight that Number rules the universe to assert the primacy of Algorithm, not Number. And those are **discrete** algorithms, it's a digital philosophy! [That's a term invented by Edward Fredkin, who has worked on related ideas.] Then Wolfram sets out to survey all possible worlds, at least all the simple ones. (That's why his book is so thick!) Along the way he finds a lot of interesting stuff. For example, Wolfram's cellular automata rule 110 is a universal computer, an amazingly simple one, that can carry out **any** computation. *A New Kind of Science* is an attempt to discover the laws of the universe by pure thought, to search systematically for God's building blocks!

The limits of reductionism: In what sense can biology and psychology be reduced to mathematics and physics?! This is indeed the acid test of a reductionist viewpoint! Historical contingency is often invoked here: life as "frozen accidents" (mutations), not something fundamental [Wolfram, Gould]. Work on artificial life (Alife) plus advances in robotics are particularly aggressive reductionist attempts. The normal way to "explain" life is evolution by natural selection, ignoring Darwin's own sexual selection and symbiotic/cooperative views of the origin of biological progress—new species—notably espoused by Lynn Margulis ("symbiogenesis"). Other problems with Darwinian gradualism: following the DNA as software paradigm, small changes in DNA software can produce big changes in organisms, and a good way to build this software is by trading useful subroutines (this is called horizontal or lateral DNA transfer). [This is how bacteria acquire immunity to antibiotics.] In fact, there is a lack of fossil evidence for many intermediate forms [already noted by Darwin], which is evidence for rapid production of new species (so-called "punctuated equilibrium").

III. WHAT DO WORKING SCIENTISTS THINK ABOUT SIMPLICITY AND COMPLEXITY?

"Science itself, therefore, may be regarded as a minimal problem, consisting of the completest possible presentment of facts with the *least possible expenditure of thought* . . . Those ideas that hold good throughout the widest domains of research and that supplement the greatest amount of experience, are *the most scientific.*"

—Ernst Mach, *The Science of Mechanics,* 1893, Chapter IV, Section IV, "The Economy of Science," reprinted in Newman, *The World of Mathematics,* 1956.

"Furthermore, the attitude that theoretical physics does not explain phenomena, but only classifies and correlates, is today accepted by most theoretical physicists. This means that the criterion of success for such a theory is simply whether it can, by a simple and elegant classifying and correlating scheme, cover very many phenomena, which without this scheme would seem complicated and heterogeneous, and whether the scheme even covers phenomena which were not considered or even not known at the time when the scheme was evolved. (These two latter statements express, of course, the unifying and the predicting power of a theory.)"

—John von Neumann, "The Mathematician," 1947, reprinted in Newman, *The World of Mathematics,* 1956, and in Bródy, Vámos, *The Neumann Compendium,* 1995.

"These fundamental concepts and postulates, which cannot be further reduced logically, form the essential part of a theory, which reason cannot touch. It is the grand object of all theory to make these irreducible elements as simple and as few in number as possible . . . [As] the distance in thought between the fundamental concepts and laws on the one side and, on the other, the conclusions which have to be brought into relation with our experience grows larger and larger, the simpler the logical structure becomes—that is to say, the smaller the number of logically independent conceptual elements which are found necessary to support the structure."

—Einstein, "On the Method of Theoretical Physics," 1934,
reprinted in Einstein, *Ideas and Opinions,* 1954.

"The aim of science is, on the one hand, a comprehension, as *complete* as possible, of the connection between the sense experiences in their totality, and, on the other hand, the accomplishment of this aim *by the use of a minimum of primary concepts and relations.* (Seeking as far as possible, logical unity in the world picture, i.e., paucity in logical elements.)

"Physics constitutes a logical system of thought which is in a state of evolution, whose basis cannot be distilled, as it were, from experience by an inductive method, but can only be arrived at by free invention . . . Evolution is proceeding in the direction of increased simplicity of the logical basis. In order further to approach this goal, we must resign to the fact that the logical basis departs more and more from the facts of experience, and that the path of our thought from the fundamental basis to those derived propositions, which correlate with sense experiences, becomes continually harder and longer."

—Einstein, "Physics and Reality," 1936,
reprinted in Einstein, *Ideas and Opinions,* 1954.

"[S]omething general will have to be said . . . about the points of view from which physical theories may be analyzed critically . . . The first point of view is obvious: the theory must not contradict empirical facts . . . The second point of view is not concerned with the relationship to the observations but with the premises of the theory itself, with what may briefly but vaguely be characterized as the 'naturalness' or 'logical simplicity' of the premises (the basic concepts and the relations between these) . . . We prize a theory more highly if, from the logical standpoint, it does not involve an arbitrary choice among theories that are equivalent and possess analogous structures . . . I must confess herewith that I cannot at this point, and perhaps not at all, replace these hints by more precise definitions. I believe, however, that a sharper formulation would be possible."

—Einstein, "Autobiographical Notes," originally published in Schilpp, *Albert Einstein, Philosopher-Scientist,* 1949, and reprinted as a separate book in 1979.

"What, then, impels us to devise theory after theory? Why do we devise theories at all? The answer to the latter question is simply: because we enjoy 'comprehending,' i.e., reducing phenomena by the process of logic to something already known or (apparently) evident. New theories are first of all necessary when we encounter new facts which cannot be 'explained' by existing theories. But this motivation for setting up new theories is, so to speak, trivial, imposed from without. There is another, more subtle motive of no less importance. This is the striving toward unification and simplification of the premises of the theory as a whole (i.e., Mach's principle of economy, interpreted as a logical principle).

"There exists a passion for comprehension, just as there exists a passion for music. That passion is rather common in children, but gets lost in most people later on. Without this passion, there would be neither mathematics nor natural science. Time and again the passion for understanding has led to the illusion that man is able to comprehend the objective world rationally, by pure thought, without any empirical foundations—in short, by metaphysics. I believe that every true theorist is a kind of tamed metaphysicist, no matter how pure a 'positivist' he may fancy himself. The metaphysicist believes that the logically simple is also the real. The tamed metaphysicist believes that not all that is logically simple is embodied in experienced reality, but that the totality of all sensory experience can be 'comprehended' on the basis of a conceptual system built on premises of great simplicity. The skeptic will say that this is a 'miracle creed.' Admittedly so, but it is a miracle creed which has been borne out to an amazing extent by the development of science."

—Einstein, "On the Generalized Theory of Gravitation," 1950, reprinted in Einstein, *Ideas and Opinions,* 1954.

"One of the most important things in this 'guess—compute consequences—compare with experiment' business is to know when you are right. It is possible to know when you are right way ahead of checking all the consequences. You can recognize truth by its beauty and simplicity. It is always easy when you have made a guess, and done two or three little calculations to make sure that it is not obviously wrong, to know that it is right. When you get it right, it is obvious that it is right—at least if you have any experience—because usually what happens is that more comes out than goes in. Your guess is, in fact, that something is very simple. If you cannot see immediately that it is wrong, and it is simpler than it was before, then it is right. The inexperienced, and crackpots, and people like that, make guesses that are simple, but you can immediately see that they are wrong, so that does not count. Others, the inexperienced students, make guesses that are very complicated, and it sort of looks as if it is all right, but I know it is not true because the truth always turns out to be simpler than you thought. What we need is imagination, but imagination in a terrible strait-jacket. We have to find a new view of the world that has to agree with everything that is known, but disagree in its predictions somewhere, otherwise it is not interesting. And in that disagreement it must agree with nature . . ."

—Feynman, *The Character of Physical Law,* 1965, Chapter 7, "Seeking New Laws."

"It is natural that a man should consider the work of his hands or his brain to be useful and important. Therefore nobody will object to an ardent experimentalist boasting of his measurements and rather looking down on the 'paper and ink' physics of his theoretical friend, who on his part is proud of his lofty ideas and despises the dirty fingers of the other. But in recent years this kind of friendly rivalry has changed into something more serious . . . [A] school of extreme experimentalists . . . has gone so far as to reject theory altogether . . . There is also a movement in the opposite direction . . . claiming that to the mind well trained in mathematics and epistemology the laws of Nature are manifest without appeal to experiment.

"Given the knowledge and the penetrating brain of our mathematician, Maxwell's equations are a result of pure thinking and the toil of experimenters antiquated and superfluous. I need hardly explain to you the fallacy of this standpoint. It lies in the fact that none of the notions used by the mathematicians, such as potential, vector potential, field vectors, Lorentz transformations, quite apart from the principle of action itself, are evident or given *a priori.* Even if an extremely gifted mathematician had constructed them to describe the properties of a possible world, neither he nor anybody else would have had the slightest idea how to apply them to the real world.

"Charles Darwin, my predecessor in my Edinburgh chair, once said something like this: 'The Ordinary Man can see a thing an inch in front of his nose; a few can see things 2 inches distant; if anyone can see it at 3 inches, he is a man of genius.' I have tried to describe to you some of the acts of these 2- or 3-inch men. My admiration of them is not diminished by the consciousness of the fact that they were guided by the experience of the whole human race to the right place into which to poke their noses. I have also not endeavoured to analyse the idea of beauty or perfection or simplicity of a natural law which has often guided the correct divination. I am convinced that such an analysis would lead to nothing; for these ideas are themselves subject to development. We learn something new from every new case, and I am not inclined to accept final theories about invariable laws of the human mind.

"My advice to those who wish to learn the art of scientific prophecy is not to rely on abstract reason, but to decipher the secret language of Nature from Nature's documents, the facts of experience."

—Max Born, *Experiment and Theory in Physics,* 1943, pp. 1, 8, 34–35, 44.

These eloquent discussions of the role that simplicity and complexity play in scientific discovery by these distinguished 20th-century scientists show the importance that they ascribe to these questions.

In my opinion, the fundamental point is this: The belief that the universe is rational, lawful, is of no value if the laws are too complicated for us to comprehend, and is even meaningless if the laws are as complicated as our observations, since the laws are then no simpler than the world they are supposed to explain. As we saw in the previous section, this was emphasized (and attributed to Leibniz) by Hermann Weyl, a fine mathematician and mathematical physicist.

But perhaps we are overemphasizing the role that the notions of simplicity and complexity play in science?

In his beautiful 1943 lecture published as a small book on *Experiment and Theory in Physics,* the theoretical physicist Max Born criticized those who think that we can understand Nature by pure thought, without hints from experiments. In particular, he was referring to now forgotten and rather fanciful theories put forth by Eddington and Milne. Now he might level these criticisms at string theory and at Stephen Wolfram's *A New Kind of Science* [Jacob T. Schwartz, private communication].

Born has a point. Perhaps the universe **is** complicated, not simple! This certainly seems to be the case in biology more than in physics. Then thought alone is insufficient; we need empirical data. But simplicity certainly reflects what we mean by understanding: **understanding is compression.** So perhaps this is more about the human mind than it is about the universe. Perhaps our emphasis on simplicity says more about us than it says about the universe!

Now we'll try to capture some of the essential features of these philosophical ideas in a mathematical theory.

IV. A MATHEMATICAL THEORY OF SIMPLICITY, COMPLEXITY AND IRREDUCIBILITY: AIT

The basic idea of algorithmic information theory (AIT) is that a scientific theory is a computer program, and the smaller, the more concise the program is, the better the theory!

But the idea is actually much broader than that. **The central idea of algorithmic information theory is reflected in the belief that the following diagrams all have something fundamental in common.** In each case, ask how much information we put in versus how much we get out. And everything is digital, discrete.

Shannon information theory (communications engineering), noiseless coding:

encoded message → **Decoder** → original message

Model of scientific method:

scientific theory → **Calculations** → empirical/experimental data

Algorithmic information theory (AIT), definition of program-size complexity:

program → **Computer** → output

Central dogma of molecular biology:

DNA → **Embryogenesis/Development** → organism

(In this connection, see Küppers, *Information and the Origin of Life,* 1990.) Turing/Post abstract formulation of a Hilbert-style formal axiomatic mathematical theory as a mechanical procedure for systematically deducing all possible consequences from the axioms:

axioms → **Deduction** → theorems

Contemporary physicists' efforts to find a Theory of Everything (TOE):

TOE → **Calculations** → Universe

Leibniz, *Discourse on Metaphysics,* 1686:

Ideas → **Mind of God** → The World

In each case the left-hand side is smaller, much smaller, than the right-hand side. In each case, the right-hand side can be constructed (reconstructed) mechanically, or systematically, from the left-hand side. And in each case we want to keep the right-hand side fixed while making the left-hand side as small as possible. Once this is accomplished, we can use the size of the left-hand side as a measure of the simplicity or the complexity of the corresponding right-hand side.

Starting with this one simple idea, of looking at the size of computer programs, or at program-size complexity, you can develop a sophisticated, elegant mathematical theory, AIT, as you can see in my four Springer-Verlag volumes listed in the bibliography of this paper.

But I must confess that AIT makes a large number of **important hidden assumptions!** What are they?

Well, one important hidden assumption of AIT is that the choice of computer or of computer programming language is not too important, that it does not affect program-size complexity too much, in any fundamental way. This is debatable.

Another important tacit assumption: we use the discrete computation approach of Turing 1936, eschewing computations with "real" (infinite-precision) numbers like $\pi = 3.1415926 \ldots$ which have an infinite number of digits when written in decimal notation, but which correspond, from a geometrical point of view, to a single point on a line, an elemental notion in continuous, but not in discrete, mathematics. Is the universe **discrete** or **continuous**? Leibniz is famous for his work on continuous mathematics. AIT sides with the discrete, not with the continuous [Françoise Chaitin-Chatelin, private communication].

Also, in AIT we completely ignore the **time** taken by a computation, concentrating only on the **size** of the program. And the computation run-times may be monstrously large, quite impracticably so, in fact, totally astronomical in size. But trying to take time into account destroys AIT, an elegant, simple theory of complexity, and one which imparts much intuitive understanding. So I think that it is a mistake to try to take time into account when thinking about this kind of complexity.

We've talked about simplicity and complexity, but what about **irreducibility**? Now let's apply AIT to mathematical logic and obtain some limitative metatheorems. However, following Turing 1936 and Post 1944, I'll use the notion of algorithm to deduce limits to formal reasoning, not Gödel's original 1931 approach. I'll take the position that a Hilbert-style mathematical theory, a formal axiomatic theory, is a mechanical procedure for systematically generating all the theorems by running through all possible proofs, systematically deducing all consequences of the axioms. [In a way, this point of view was anticipated by Leibniz with his *lingua characteristica universalis.*] Consider the size in bits of the algorithm for doing this. This is how we measure the simplicity or complexity of the formal axiomatic theory. It's just another instance of program-size complexity!

But at this point, Chaitin-Chatelin insists, I should admit that we are making an extremely embarrassing hidden assumption, which is that you can systematically run through all the proofs. This assumption, which is bundled into my definition of a formal axiomatic theory, means that we are assuming that the language of our theory is static, and that no new concepts can ever emerge. But no human language or field of thought is static! (And computer programming languages aren't static either, which can be quite a nuisance.) And this idea of being able to make a numbered list with all possible proofs was clearly anticipated by Émile Borel in 1927 when he pointed out that there is a real number with the problematical property that its *N*th digit after the decimal point gives us the answer to the *N*th yes/no question in French. (Borel's work was brought to my attention by Vladimir Tasić in his book *Mathematics and the Roots of Postmodern Thought,* 2001, where he points out that in some ways it anticipates the Ω number that I'll discuss in Section IX. Borel's paper is reprinted in Mancosu, *From Brouwer to Hilbert,* 1998, pp. 296–300.)

Yes, I agree, a Hilbert-style formal axiomatic theory is indeed a fantasy, but it is a fantasy that inspired many people, and one that even helped to lead to the creation of modern programming languages. It is a fantasy that it is useful to take seriously long enough for us to show in Section VI that even if you are willing to accept all these tacit assumptions, something else is terribly wrong. Formal axiomatic theories can be criticized from within, as well as from without. And it is far from clear how weakening these tacit assumptions would make it easier to prove the irreducible mathematical truths that are exhibited in Section VI.

And the idea of a fixed, static computer programming language in which you write the computer programs whose size you measure is also a fantasy. Real computer programming languages don't stand still, they evolve, and the size of the computer program you need to perform a given task can therefore change. Mathematical models of the world like these are always approximations, "lies that help us to see the truth" (Picasso). Nevertheless, if done properly, they can impart insight and understanding, they can help us to comprehend, they can reveal unexpected connections . . .

V. FROM COMPUTATIONAL IRREDUCIBILITY TO LOGICAL IRREDUCIBILITY. EXAMPLES OF COMPUTATIONAL IRREDUCIBILITY: "ELEGANT" PROGRAMS

Our goal in this section and the next is to use AIT to establish the existence of **irreducible mathematical truths.** What are they, and why are they important?

Following Euclid's *Elements,* a mathematical truth is established by reducing it to simpler truths until self-evident truths—"axioms" or "postulates" [atoms of thought!]—are reached. Here we exhibit an extremely large class of mathematical truths that are not at all self-evident but which are **not** consequences of any principles simpler than they are.

Irreducible truths are highly problematical for traditional philosophies of mathematics, but as discussed in Section VIII, they can be accommodated in an emerging "quasi-empirical" school of the foundations of mathematics, which says that physics and mathematics are not that different.

Our path to logical irreducibility starts with computational irreducibility. Let's start by calling a computer program "elegant" if no smaller program in the same language produces exactly the same output. There are lots of elegant programs, at least one for each output. And it doesn't matter how **slow** an elegant program is, all that matters is that it be as **small** as possible.

An elegant program viewed as an object in its own right is computationally irreducible. Why? Because otherwise you can get a more concise program for its output by computing it first and then running it. Look at this diagram:

$$\text{program}_2 \longrightarrow \textbf{Computer} \longrightarrow \text{program}_1 \longrightarrow \textbf{Computer} \longrightarrow \text{output}$$

If program_1 is as concise as possible, then program_2 cannot be much more concise than program_1. Why? Well, consider a fixed-sized routine for running a program and then immediately running its output. Then

$$\text{program}_2 + \text{fixed-size routine} \longrightarrow \textbf{Computer} \longrightarrow \text{output}$$

produces exactly the same output as program_1 and would be a more concise program for producing that output than program_1 is. But this is impossible because it contradicts our hypothesis that program_1 was already as small as possible. *Q.E.D.*

Why should elegant programs interest philosophers? Well, because of Occam's razor, because the best theory to explain a fixed set of data is an elegant program!

But how can we get irreducible truths? Well, just try **proving** that a program is elegant!

VI. IRREDUCIBLE MATHEMATICAL TRUTHS. EXAMPLES OF LOGICAL IRREDUCIBILITY: PROVING A PROGRAM IS ELEGANT

Hauptsatz: *You cannot prove that a program is elegant if its size is substantially larger than the size of the algorithm for generating all the theorems in your theory.*

Proof: The basic idea is to run the first provably elegant program you encounter when you systematically generate all the theorems, and that is substantially larger than the size of the algorithm for generating all the theorems. Contradiction, unless no such theorem can be demonstrated, or unless the theorem is false.

Now I'll explain why this works. We are given a formal axiomatic mathematical theory:

theory = program ⟶ **Computer** ⟶ set of all theorems

We may suppose that this theory is an elegant program, i.e., as concise as possible for producing the set of theorems that it does. Then the size of this program is by definition the complexity of the theory, since it is the size of the smallest program for systematically generating the set of all the theorems, which are all the consequences of the axioms. Now consider a fixed-size routine with the property that

theory + fixed-size routine ⟶ **Computer** ⟶
output of the first provably elegant program larger than
complexity of theory

More precisely,

theory + fixed-size routine ⟶ **Computer** ⟶
output of the first provably elegant program larger than
(complexity of theory + size of the fixed-size routine)

This proves our assertion that a mathematical theory cannot prove that a program is elegant if that program is substantially larger than the complexity of the theory.

Here is the proof of this result in more detail. The fixed-size routine knows its own size and is given the theory, a computer program for generating theorems, whose size it measures and which it then runs, until the first theorem is encountered asserting that a particular program *P* is elegant that is larger than the total input to the computer. The fixed-size routine then runs the program *P*, and finally produces as output the same output as *P* produces. But this is impossible, because the output from *P* cannot be obtained from a program that is smaller than *P* is, not if, as we assume by hypothesis, all the theorems of the theory are true and *P* is actually elegant. Therefore *P* cannot exist. In other words, if there is a provably elegant program *P* whose size is greater than the complexity of the theory + the size of this fixed-size routine, either *P* is actually inelegant or we have a contradiction. *Q.E.D.*

Because no mathematical theory of finite complexity can enable you to determine all the elegant programs, the following is immediate:

Corollary: *The mathematical universe has infinite complexity.*

[On the other hand, our current mathematical theories are **not** very complex. On pages 773–774 of *A New Kind of Science,* Wolfram makes this point by exhibiting essentially all of the axioms for traditional mathematics—in just two pages! However, a program to generate all the theorems would be larger.]

This strengthens Gödel's 1931 refutation of Hilbert's belief that a single, fixed formal axiomatic theory could capture all of mathematical truth.

Given the significance of this conclusion, it is natural to demand more information. You'll notice that I never said **which** computer programming language I was using!

Well, you can actually carry out this proof using either high-level languages such as the version of LISP that I use in *The Unknowable,* or using low-level binary machine languages, such as the one that I use in *The Limits of Mathematics.* In the case of a high-level computer programming language, one measures the size of a program in characters (or 8-bit bytes) of text. In the case of a binary machine language, one measures the size of a program in 0/1 bits. My proof works either way.

But I must confess that not all programming languages permit my proof to work out this neatly. The ones that do are the kinds of programming languages that you use in AIT, the ones for which program-size complexity has elegant properties instead of messy ones, the ones that directly expose the fundamental nature of this complexity concept (which is also called algorithmic information content), not the programming languages that bury the basic idea in a mass of messy technical details.

This paper started with philosophy, and then we developed a mathematical theory. Now let's go back to philosophy. In the last three sections of this paper we'll discuss the philosophical implications of AIT.

VII. COULD WE EVER BE SURE THAT WE HAD THE ULTIMATE TOE? [BARROW 1995]

> "The search for a 'Theory of Everything' is the quest for an ultimate compression of the world. Interestingly, Chaitin's proof of Gödel's incompleteness theorem using the concepts of complexity and compression reveals that Gödel's theorem is equivalent to the fact that one cannot prove a sequence to be incompressible. We can never prove a compression to be the ultimate one; there might be a yet deeper and simpler unification waiting to be found."
>
> —John Barrow, essay on "Theories of Everything" in Cornwell, *Nature's Imagination,* 1995, reprinted in Barrow, *Between Inner Space and Outer Space,* 1999.

Here is the first philosophical application of AIT. According to astrophysicist John Barrow, my work implies that even if we had the optimum, perfect, minimal (elegant!) TOE, we could never be sure a simpler theory would not have the same explanatory power.

("Explanatory power" is a pregnant phrase, and one can make a case that it is a better name to use than the dangerous word "complexity," which has many other possible meanings. One could then speak of a theory with *N* bits of algorithmic explanatory power, rather than describe it as a theory having a program-size complexity of *N* bits [Françoise Chaitin-Chatelin, private communication]).

Well, you can dismiss Barrow by saying that the idea of having the ultimate TOE is pretty crazy—who expects to be able to read the mind of God?! Actually, Wolfram believes that a systematic computer search might well find the ultimate TOE. [See pages 465–471, 1024–1027 of *A New Kind of Science.*] I hope he continues working on this project!

In fact, Wolfram thinks that he not only might be able to find the ultimate TOE, he might even be able to show that it is the simplest possible TOE! How does he escape the impact of my results? Why doesn't Barrow's observation apply here?

First of all, Wolfram is not very interested in proofs, he prefers computational evidence. Second, Wolfram does not use program-size complexity as his complexity measure. He uses much more down-to-earth complexity measures. Third, he is concerned with extremely simple systems, while my methods apply best to objects with high complexity.

Perhaps the best way to explain the difference is to say that he is looking at "hardware" complexity, and I'm looking at "software" complexity. The objects he studies have complexity less than or equal to that of a universal computer. Those I study have complexity much larger than a universal computer. For Wolfram, a universal computer is the maximum possible complexity, and for me it is the minimum possible complexity.

Anyway, now let's see what's the message from AIT for the working mathematician.

VIII. SHOULD MATHEMATICS BE MORE LIKE PHYSICS? MUST MATHEMATICAL AXIOMS BE SELF-EVIDENT?

"A deep but easily understandable problem about prime numbers is used in the following to illustrate the parallelism between the heuristic reasoning of the mathematician and the inductive reasoning of the physicist . . . [M]athematicians and physicists think alike; they are led, and sometimes misled, by the same patterns of plausible reasoning."

—George Pólya, "Heuristic Reasoning in the Theory of Numbers," 1959, reprinted in Alexanderson, *The Random Walks of George Pólya,* 2000.

"The role of heuristic arguments has not been acknowledged in the philosophy of mathematics, despite the crucial role that they play in mathematical discovery. The mathematical notion of proof is strikingly at variance with the notion of proof in other areas . . . Proofs given by physicists do admit degrees: of two proofs given of the same assertion of physics, one may be judged to be more correct than the other."

—Gian-Carlo Rota, "The Phenomenology of Mathematical Proof," 1997, reprinted in Jacquette, *Philosophy of Mathematics,* 2002, and in Rota, *Indiscrete Thoughts,* 1997.

"There are two kinds of ways of looking at mathematics . . . the Babylonian tradition and the Greek tradition . . . Euclid discovered that there was a way in which all the theorems of geometry could be ordered from a set of axioms that were particularly simple . . . The Babylonian attitude . . . is that you know all of the various theorems and many of the connections in between, but you have never fully realized that it could all come up from a bunch of axioms . . . [E]ven in mathematics you can start in different places . . . In physics we need the Babylonian method, and not the Euclidian or Greek method."

—Richard Feynman, *The Character of Physical Law,* 1965, Chapter 2, "The Relation of Mathematics to Physics."

"The physicist rightly dreads precise argument, since an argument which is only convincing if precise loses all its force if the assumptions upon which it is based are slightly changed, while an argument which is convincing though imprecise may well be stable under small perturbations of its underlying axioms."

—Jacob Schwartz, "The Pernicious Influence of Mathematics on Science," 1960, reprinted in Kac, Rota, Schwartz, *Discrete Thoughts,* 1992.

"It is impossible to discuss realism in logic without drawing in the empirical sciences . . . A truly realistic mathematics should be conceived, in line with physics, as a branch of the theoretical construction of the one real world and should adopt the same sober and cautious attitude toward hypothetic extensions of its foundation as is exhibited by physics."

—Hermann Weyl, *Philosophy of Mathematics and Natural Science,* 1949, Appendix A, "Structure of Mathematics," p. 235.

The above quotations are eloquent testimonials to the fact that although mathematics and physics are different, maybe they are not **that** different! Admittedly, math organizes our mathematical experience, which is mental or computational, and physics organizes our physical experience. [And in physics everything is an approximation, no equation is exact.] They are certainly not exactly the same, but maybe it's a matter of degree, a continuum of possibilities, and not an absolute, black-and-white difference.

Certainly, as both fields are currently practiced, there is a definite difference in **style.** But that could change, and is to a certain extent a matter of fashion, not a fundamental difference.

A good source of essays that I—but perhaps not the authors!—regard as generally supportive of the position that math be considered a branch of physics is Tymoczko, *New Directions in the Philosophy of Mathematics,* 1998. In particular there you will find an essay by Lakatos giving the name "quasi-empirical" to this view of the nature of the mathematical enterprise.

Why is my position on math "quasi-empirical"? Because, as far as I can see, this is the only way to accommodate the existence of irreducible mathematical facts gracefully. Physical postulates are never self-evident, they are justified pragmatically, and so are close relatives of the not at all self-evident irreducible mathematical facts that I exhibited in Section VI.

I'm not proposing that math is a branch of physics just to be controversial. I was forced to do this against my will! This happened in spite of the fact that I'm a mathematician and I love mathematics, and in spite of the fact that I started with the traditional Platonist position shared by most working mathematicians. I'm proposing this because I want mathematics to work better and be more productive. Proofs are fine, but if you can't find a proof, you should go ahead using heuristic arguments and conjectures.

Wolfram's *A New Kind of Science* also supports an experimental, quasi-empirical way of doing mathematics. This is partly because Wolfram is a physicist, partly because he believes that unprovable truths are the rule, not the exception, and partly because he believes that our current mathematical theories are highly arbitrary and contingent. Indeed, his book may be regarded as a very large chapter in experimental math. In fact, he had to develop his own programming language, *Mathematica,* to be able to do the massive computations that led him to his conjectures.

See also Tasić, *Mathematics and the Roots of Postmodern Thought,* 2001, for an interesting perspective on intuition versus formalism. This is a key question—indeed in my opinion it's an inescapable issue—in any discussion of how the game of mathematics should be played. And it's a question with which I, as a working mathematician, am passionately concerned, because, as we discussed in Section VI, formalism has severe limitations. Only intuition can enable us to go forward and create new ideas and more powerful formalisms.

And what are the wellsprings of mathematical intuition and creativity? In his important forthcoming book on creativity, Tor Nørretranders makes the case that a peacock, an elegant, graceful woman, and a beautiful mathematical theory are all shaped by the same forces, namely what Darwin referred to as "sexual selection." Hopefully this book will be available soon in a language other than Danish! Meanwhile, see my dialogue with him in my book *Conversations with a Mathematician.*

Now, for our last topic, let's look at the entire physical universe!

IX. IS THE UNIVERSE LIKE π OR LIKE Ω? REASON VERSUS RANDOMNESS! [BRISSON, MEYERSTEIN 1995]

"Parce qu'on manquait d'une définition rigoreuse de complexité, celle qu'a proposée la TAI [théorie algorithmique de l'information], confondre π avec Ω a été plutôt la règle que l'exception. Croire, parce que nous avons ici affaire à une croyance, que toutes les suites, puisqu'elles ne sont que l'enchaînement selon une règle rigoureuse de symboles déterminés, peuvent toujours être comprimées en quelque chose de plus simple, voilà la source de l'erreur du réductionnisme. Admettre la complexité a toujours paru insupportable aux philosophes, car c'était renoncer à trouver un sens rationnel à la vie des hommes."

—Brisson, Meyerstein, *Puissance et Limites de la Raison,* 1995, "Postface. L'erreur du réductionnisme," p. 229.

First let me explain what the number Ω is. It's the jewel in AIT's crown, and it's a number that has attracted a great deal of attention, because it's a very **dangerous** number! Ω is defined to be the halting probability of what computer scientists call a universal computer, or universal Turing machine. (In fact, the precise value of Ω actually depends on the choice of computer, and in *The Limits of Mathematics* I've done that, I've picked one out.) So Ω is a probability and therefore it's a real number, a number measured with infinite precision, that's between zero and one. That may not sound too dangerous!

[It's ironic that the star of a discrete theory is a real number! This illustrates the creative tension between the continuous and the discrete.]

What's dangerous about Ω is that (a) it has a simple, straightforward mathematical definition, but at the same time (b) its numerical value is maximally unknowable, because a formal mathematical theory whose program-size complexity or explanatory power is N bits cannot enable you to determine more than N bits of the base-two expansion of Ω! In other words, if you want to calculate Ω, theories don't help very much, since it takes N bits of theory to get N bits of Ω. In fact, the base-two bits of Ω are maximally complex, there's no redundancy, and Ω is the prime example of how unadulterated infinite complexity arises in pure mathematics!

How about $\pi = 3.1415926 \ldots$ the ratio of the circumference of a circle to its diameter? Well, π **looks** pretty complicated, pretty lawless. For example, all its digits seem to be equally likely, although this has never been proven. (Amazingly enough, there's been some recent progress in this direction by Bailey and Crandall.) [In any base all the digits of Ω are equally likely. This is called "Borel normality." For a proof, see my book *Exploring Randomness.* For the latest on Ω, see Calude, *Information and Randomness.*] If you are given a bunch of digits from deep inside the decimal expansion of π, and you aren't told where they come from, there doesn't seem to be any redundancy, any pattern. But of course, according to AIT, π in fact only has **finite** complexity, because there are algorithms for calculating it with arbitrary precision. (In fact, some terrific new ways to calculate π have been discovered by Bailey, Borwein and Plouffe. π lives, it's not a dead subject!)

Following Brisson, Meyerstein, *Puissance et Limites de la Raison,* 1995, let's now finally discuss whether the physical universe is like π = 3.1415926 … which only has a finite complexity, namely the size of the smallest program to generate π, or like Ω, which has unadulterated infinite complexity. Which is it?!

Well, if you believe in quantum physics, then Nature plays dice, and that generates complexity, an infinite amount of it, for example, as frozen accidents, mutations that are preserved in our DNA. So at this time most scientists would bet that the universe has infinite complexity, like Ω does. But then the world is incomprehensible, or at least a large part of it will always remain so, the accidental part, all those frozen accidents, the contingent part.

But some people still hope that the world has finite complexity like π, it just **looks** like it has high complexity. If so, then we might eventually be able to comprehend everything, and there is an ultimate TOE! But then you have to believe that quantum mechanics is wrong, as currently practiced, and that all that quantum randomness is really only **pseudo-randomness,** like what you find in the digits of π. You have to believe that the world is actually deterministic, even though our current scientific theories say that it isn't!

I think Vienna physicist Karl Svozil feels that way [private communication; see his *Randomness & Undecidability in Physics,* 1994]. I know Stephen Wolfram does, he says so in his book. Just take a look at the discussion of fluid turbulence and of the second law of thermodynamics in *A New Kind of Science.* Wolfram believes that very simple deterministic algorithms ultimately account for all the apparent complexity we see around us, just like they do in π. He believes that the world **looks** very complicated, but is actually very simple. There's no randomness, there's only pseudo-randomness. Then nothing is contingent, everything is necessary, everything happens for a reason. [Leibniz!]

[In fact, Wolfram himself explicitly makes the connection with π. See **meaning of the universe** on page 1027 of *A New Kind of Science.*]

Who knows! Time will tell!

Or perhaps from **inside** this world we will never be able to tell the difference, only an **outside** observer could do that [Svozil, private communication].

POSTSCRIPT

Readers of this paper may enjoy the somewhat different perspective in my chapter "Complexité, logique et hasard" in Benkirane, *La Complexité.* Leibniz is there too. In addition, see my *Conversations with a Mathematician,* a book on philosophy disguised as a series of dialogues—not the first time that this has happened!

Last but not least, see Zwirn, *Les Limites de la Connaissance,* that also supports the thesis that understanding is compression, and the masterful multi-author two-volume work, *Kurt Gödel, Wahrheit & Beweisbarkeit,* a treasure trove of information about Gödel's life and work.

ACKNOWLEDGMENT

Thanks to Tor Nørretranders for providing the original German for the Einstein quotation at the beginning of this paper, and also the word-for-word translation.

The author is grateful to Françoise Chaitin-Chatelin for innumerable stimulating philosophical discussions. He dedicates this paper to her unending quest to understand.

BIBLIOGRAPHY

- Gerald W. Alexanderson, *The Random Walks of George Pólya,* MAA, 2000.
- John D. Barrow, Frank J. Tipler, *The Anthropic Cosmological Principle,* Oxford University Press, 1986.
- John D. Barrow, *Between Inner Space and Outer Space,* Oxford University Press, 1999.
- Reda Benkirane, *La Complexité, Vertiges et Promesses,* Le Pommier, 2002.
- Max Born, *Experiment and Theory in Physics,* Cambridge University Press, 1943. Reprinted by Dover, 1956.
- Luc Brisson, F. Walter Meyerstein, *Inventer l'Univers,* Les Belles Lettres, 1991.
- Luc Brisson, F. Walter Meyerstein, *Inventing the Universe,* SUNY Press, 1995.
- Luc Brisson, F. Walter Meyerstein, *Puissance et Limites de la Raison,* Les Belles Lettres, 1995.
- F. Bródy, T. Vámos, *The Neumann Compendium,* World Scientific, 1995.
- Bernd Buldt et al., *Kurt Gödel, Wahrheit & Beweisbarkeit. Band 2: Kompendium zum Werk,* öbv & hpt, 2002.
- Cristian S. Calude, *Information and Randomness,* Springer-Verlag, 2002.
- Gregory J. Chaitin, *The Limits of Mathematics, The Unknowable, Exploring Randomness, Conversations with a Mathematician,* Springer-Verlag, 1998, 1999, 2001, 2002.
- John Cornwell, *Nature's Imagination,* Oxford University Press, 1995.
- COSRIMS, *The Mathematical Sciences,* MIT Press, 1969.

- Albert Einstein, *Ideas and Opinions,* Crown, 1954. Reprinted by Modern Library, 1994.
- Albert Einstein, *Autobiographical Notes,* Open Court, 1979.
- Richard Feynman, *The Character of Physical Law,* MIT Press, 1965. Reprinted by Modern Library, 1994, with a thoughtful introduction by James Gleick.
- Richard P. Feynman, Robert B. Leighton, Matthew Sands, *The Feynman Lectures on Physics,* Addison-Wesley, 1963.
- Dale Jacquette, *Philosophy of Mathematics*, Blackwell, 2002.
- Mark Kac, Gian-Carlo Rota, Jacob T. Schwartz, *Discrete Thoughts,* Birkhäuser, 1992.
- Eckehart Köhler et al., *Kurt Gödel, Wahrheit & Beweisbarkeit. Band 1: Dokumente und historische Analysen,* öbv & hpt, 2002.
- Bernd-Olaf Küppers, *Information and the Origin of Life,* MIT Press, 1990.
- G. W. Leibniz, *Philosophical Essays,* edited and translated by Roger Ariew and Daniel Garber, Hackett, 1989.
- Ernst Mach, *The Science of Mechanics,* Open Court, 1893.
- Paolo Mancosu, *From Brouwer to Hilbert,* Oxford University Press, 1998.
- Karl Menger, *Reminiscences of the Vienna Circle and the Mathematical Colloquium,* Kluwer, 1994.
- James R. Newman, *The World of Mathematics,* Simon and Schuster, 1956. Reprinted by Dover, 2000.
- Karl R. Popper, *The Logic of Scientific Discovery,* Hutchinson Education, 1959. Reprinted by Routledge, 1992.
- Gian-Carlo Rota, *Indiscrete Thoughts,* Birkhäuser, 1997.
- Paul Arthur Schilpp, *Albert Einstein, Philosopher-Scientist,* Open Court, 1949.
- Karl Svozil, *Randomness & Undecidability in Physics,* World Scientific, 1994.
- Vladimir Tasić, *Mathematics and the Roots of Postmodern Thought,* Oxford University Press, 2001.
- Thomas Tymoczko, *New Directions in the Philosophy of Mathematics,* Princeton University Press, 1998.

- Hermann Weyl, *The Open World,* Yale University Press, 1932. Reprinted by Ox Bow Press, 1989.
- Hermann Weyl, *Philosophy of Mathematics and Natural Science,* Princeton University Press, 1949.
- Stephen Wolfram, *A New Kind of Science,* Wolfram Media, 2002.
- Hervé Zwirn, *Les Limites de la Connaissance,* Odile Jacob, 2000.

INDEX

Grateful acknowledgment is made to the following for permission to reprint previously published material:

American Scientist: "Computers, Paradoxes and the Foundations of Mathematics" by Gregory J. Chaitin from *American Scientist* (March–April 2002) 90164-171. Reprinted by permission of *American Scientist.*

Robert M. Chute: Poem "Reading a Note in the Journal Nature I Learn" by Robert M. Chute, originally published in *The Beloit Poetry Journal.* Reprinted by permission of the author.

Marion D. Cohen: Poem "Math Poem" from *Crossing the Equal Sign* by Marion D. Cohen. Reprinted by permission of the author.

Crown Publishers: Excerpts from *Ideas and Opinions* by Albert Einstein. Copyright 1954 and renewed 1982 by Crown Publishers, Inc. Reprinted by permission of Crown Publishers, a division of Random House, Inc.

The MIT Press: Excerpts from *The Character of Physical Law* by Richard Feyman. Reprinted by The MIT Press.

Ox Bow Press: Excerpt from *The Open World* by Hermann Weyl. Reprinted courtesy of Ox Bow Press, Woodbridge, CT.

The University of Chicago Press: Excerpt from "The Mathematician" by J. von Neumann from *The Works of the Mind* edited by Heywood & Neff. Reprinted by permission of The University of Chicago Press.

Viking Penguin: Excerpts from "The Rose of Paracelsus" from *Collected Fictions* by Jorge Luis Borges, translated by Andrew Hurley. Copyright © 1998 by Maria Kodama. Translation copyright © 1998 by Penguin Putnam Inc. Reprinted by permission of Viking Penguin, a division of Penguin Group (USA) Inc.

ABOUT THE AUTHOR

Gregory Chaitin has devoted his life to the attempt to understand what mathematics can and cannot achieve, and is a member of the digital philosophy/digital physics movement. Its members believe that the world is built out of digital information, out of 0 and 1 bits, and they view the universe as a giant information-processing machine, a giant digital computer. In this book on the history of ideas, Chaitin traces digital philosophy back to the nearly-forgotten 17th century genius Leibniz. He also tells us how he discovered the celebrated Omega number, which marks the current boundary of what mathematics can achieve. This book is an opportunity to get inside the head of a creative mathematician and see what makes him tick, and opens a window for its readers onto a glittering world of high-altitude thought that few intellectual mountain climbers can ever glimpse.

A NOTE ON THE TYPE

The text of this book was set in a typeface called Times New Roman, designed by Stanley Morison (1889–1967) for *The Times* (London) and first introduced by that newspaper in 1932. Among typographers and designers of the twentieth century, Stanley Morison was a strong forming influence—as a typographical adviser to the Monotype Corporation, as a director of two distinguished publishing houses, and as a writer of sensibility, erudition, and keen practical sense.

Composed by Creative Graphics, Allentown, Pennsylvania
Printed and bound by Berryville Graphics, Berryville, Virginia
Designed by M. Kristen Bearse